Biological Wastewater
Treatment Systems

Biological Wastewater Treatment Systems

Theory and Operation

N. J. Horan
Department of Civil Engineering
University of Leeds
Leeds, U.K.

JOHN WILEY & SONS
Chichester · New York · Brisbane · Toronto · Singapore

Other Wiley Editorial Offices

John Wiley & Sons, Inc., 605 Third Avenue,
New York, NY 10158–0012, USA

Jacaranda Wiley Ltd, G.P.O. Box 859, Brisbane,
Queensland 4001, Australia

John Wiley & Sons (Canada) Ltd, 22 Worcester Road,
Rexdale, Ontario M9W 1L1, Canada

John Wiley & Sons (SEA) Pte Ltd, 37 Jalan Pemimpin 05–04,
Block B, Union Industrial Building, Singapore 2057

Library of Congress Cataloging-in-Publication Data:
Horan, N. J.
  Biological wastewater treatment systems : theory and operation /
N. J. Horan
    p. cm.
  Includes bibliographical references.
  ISBN 0 471 92258 7
  ISBN 0 471 92425 3 (pbk.)
  1. Sewage—Purification—Biological treatment. I. Title.
TD755.H66 1989
628.3′5—dc20   89-16629
                                                                          CIP

British Library Cataloguing in Publication Data:
Horan, N. J.
  Biological wastewater treatment systems.
  1. Waste water. Treatment. Use of organs
  I. Title
  628.3′51

ISBN 0 471 92258 7
ISBN 0 471 92425 3 pbk

Typeset by Thomson Press (India) Ltd, New Delhi
Printed in Great Britain by Redwood Books, Trowbridge, Wiltshire

Contents

Preface

The increasing concern which is being voiced as to the destruction and pollution of our environment has produced a growing worldwide awareness of the need for more effective sewage treatment. Within the UK proposals for treating and disposing of sewage and sewage sludges can produce fierce debates within the community, often resulting in the formation of pressure groups to counter unpopular or controversial proposals. In addition the contribution of sewage, treated effluents and sewage sludges to the spread of many types of human and animal infection is now being quantified. This has emphasised the vital need for improved water supply and sanitation in developing countries.

The design and operation of sewage treatment works are thus no longer the exclusive domain of the engineer and chemist; multidisciplinary teams of engineers and scientists are required in order to maximise the benefits to the community which should accrue from the installation of sewerage and sewage treatment.

This book attempts to bring together the basic information from a range of disciplines and present it in a format which can be assimilated by engineers with no biological background, and biologists who have little understanding of the processes employed for the treatment of sewage. Although it was originally intended for undergraduate students of civil and environmental engineering, it is hoped that it may prove useful to students on biological science courses which offer environment-related options. With the increased emphasis being given to the biological aspects of sewage treatment, it is also hoped that the book will be read by practising engineers and consultants, and act as a painless introduction to this approach.

Numerous people have contributed to the painful birth and slow growth of this book. The process has been made easier by the help and encouragement of my colleagues at Leeds, Duncan Mara and Ed Stentiford. They have provided a friendly working environment, together with constructive criticism and generous help and advice, whenever it was needed. Many of the ideas in this book resulted from a three year period of close involvement with the Davyhulme ETW, in Manchester, England. I would like to thank John Dolan and Mike O'Neill for the hours which were spent discussing the problems of operation and control at one of the largest sewage treatment plants in

Europe. Finally, most of the hard work on this book was done by a number of typists, illustrators and assistants, in particular Liz Baldwin who translated my shaky pencil sketches into recognisable drawings and Zena Hickinson who provided excellent administrative support in the compilation and assembly of the manuscript.

WATER

If I were called in
To construct a religion
I should make use of water.

Going to church
Would entail a fording
To dry, different clothes;

My liturgy would employ
Images of sousing,
A furious devout drench,

And I should raise in the east
A glass of water
Where any-angled light
Would congregate endlessly.

Philip Larkin

(*Reprinted by permission of Faber and Faber Ltd from* The Whitsun Weddings
by Philip Larkin.)

For Linda and Kate

1 Wastewater Characteristics and the Effects of its Discharge on Receiving Waters

1.1 THE NECESSITY FOR WASTEWATER CHARACTERISATION

A supply of clean water is an essential requirement for the establishment and maintenance of a healthy community. It acts not only as a source of potable water, but also provides valuable food supplements through supporting the growth of aquatic life, and also by its usage for irrigation in agriculture. Yet despite its importance, watercourses throughout the world are generally badly treated and frequently used to carry away the waste products of the community in the form of wastewater discharges. Water which has been utilised and discharged from domestic dwellings, institutions and commercial establishments (known as domestic wastewater), together with water discharged from manufacturing industries (known as industrial wastewater), contains a large number of potentially harmful compounds. Consequently, if it were discharged directly into a watercourse, serious damage might result to the many forms of life which inhabit this water. In addition watercourses utilised by Man, either as a source of potable water or for washing or bathing, would present potential risks of the transmission of a large number of water-related diseases.

To ensure that such problems are avoided or minimised, attention should be paid to the management of our aquatic resources and also of the pollutants which enter them. A sensible management strategy will inevitably involve analysis of the composition of wastewaters and their receiving waters. Owing to the complex nature of sampling and analysis procedures, provision of this information may prove to be an expensive and difficult task. Its planning, therefore, cannot be left solely to the experience and intuition of the engineer in the field. The initial step in the implementation of sampling and analysis programmes is to define clearly both their short- and long-term aims. This in turn will require identification of the particular problem for which a solution is sought. Typical problems associated with domestic and industrial wastewater discharges include:

1. Assessment of its pollution potential in order to establish the likely effects of its discharge on receiving water quality.

2. The type and degree of treatment which would be required in order to render the discharge harmless.

3. Identification of the source of a discharge which might be contaminating potable and bathing water supplies with human pathogens.

4. Assessment of strength and flow rates in order to levy the discharger with an appropriate treatment charge.

5. Cost-saving exercises carried out by industry with the aim of reducing water requirements and decreasing the pollution load of their effluents.

Where facilities and money are not a limiting factor, water-quality monitoring programmes may be carried out regularly in order to establish baseline standards for a given watercourse. Such information proves invaluable in assessing changes in the aquatic community, should changes in human settlement result in new wastewater discharges to a previously unpolluted environment.

Having defined the objectives of the analysis programmes and evaluation procedures, it is necessary to select the appropriate determinands of analysis; this selection should be based solely on satisfying the stated objectives.

1.2 CHARACTERISATION OF THE ORGANIC COMPONENTS OF WASTEWATER

There are many compounds and microorganisms present in a wastewater which are capable of causing the pollution of a watercourse, and pollution may be manifested in numerous ways. Because of this, a complete chemical and microbial analysis of wastewater is never performed; instead, the components which comprise a wastewater are defined in terms of three broad categories, namely organic material, inorganic material and microbial content.

Such a system of classification provides a simple framework which allows an engineer to predict adequately the pollution potential of a given wastewater. Although the division into three categories for the characterisation of pollution may seem somewhat arbitrary, the mode of pollution effected by each group is quite unique, as will be seen later in this chapter. In addition each category requires a specialised form of treatment in order to render it harmless.

Organic material

The organic components of wastewater comprise a large number of compounds which all have in common the possession of at least one carbon atom (and are thus known as carbonaceous compounds). These carbon atoms may be oxidised both chemically and biologically to yield carbon dioxide. This reaction results in a net yield of energy, and consequently carbonaceous compounds are oxidised by the majority of microorganisms to provide the energy necessary for their growth. A typical, simple carbonaceous compound is glucose and this is oxidised according to the equation:

$$C_6H_{12}O_6 + 6O_2 \longrightarrow 6CO_2 + 6O_2 \qquad (1.1)$$

Oxidation reactions of this kind may be carried out both microbially and by use of chemical oxidising agents. They are exploited as non-specific tests to indicate the amount of organic material present in a wastewater. If biological oxidation is employed the test is termed the biochemical oxygen demand (BOD), whereas for chemical oxidation the terms chemical oxygen demand (COD), permanganate value (PV) or the total oxygen demand (TOD) are used, depending on the chemical oxidising agent employed and the nature of the oxidising conditions.

Biochemical oxygen demand

This test exploits the ability of microorganisms to oxidise organic material to carbon dioxide and water using molecular oxygen as an

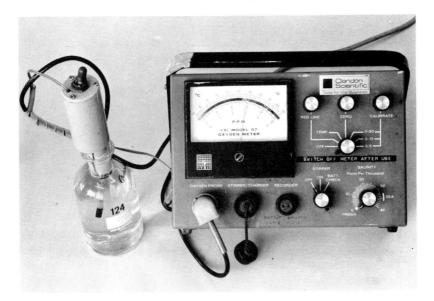

Figure 1.1 Typical laboratory apparatus for the measurement of BOD.

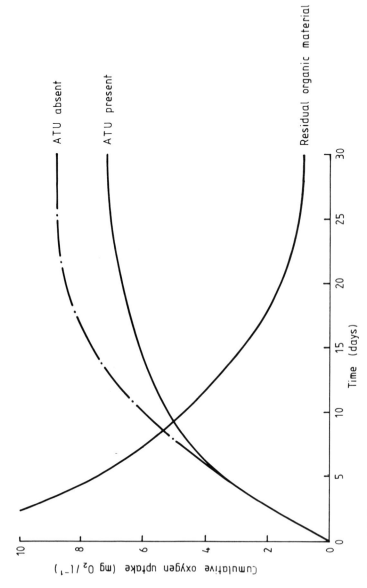

Figure 1.2 Cumulative oxygen uptake curve for a wastewater sample incubated at 20 °C in the presence and absence of allyl thiourea (ATU).

oxidising agent. The reaction takes place in a closed vessel and consequently the amount of oxygen utilised during the course of the reaction may be easily measured. A typical BOD apparatus of 300 ml volume is illustrated in Figure 1.1. A sample of the wastewater to be tested (the volume of which will be dependent upon the likely organic content) is introduced into the BOD bottle, and a 'seed', comprising a large number of microorganisms which are capable of oxidising the waste being tested, is added. Finally the bottle is filled to the top with a dilution solution which contains phosphate, $MgSO_4$, $CaCl_2$ and $FeCl_3$. The dilution solution has previously been aerated for several hours and consequently the contents of the BOD bottle are fully saturated with dissolved oxygen. The bottle is now tightly stoppered to ensure that no oxygen enters from the atmosphere, and placed in an incubator, in the dark, at 20 °C. In addition a blank is required which contains seed and dilution water only, so that corrections for the oxygen demand of the seed can be made. Changes in the dissolved oxygen content of the bottles are made at required intervals by means of a suitable oxygen electrode. If the daily cumulative oxygen uptake is followed over a 30-day time period, the curve illustrated in Figure 1.2 will result. The activity of the growing seed of microorganisms within the BOD bottle results in the rapid oxidation of the readily degradable compounds present in the waste, until the point at which they are all depleted. This point generally occurs within the first 7 days. The process of carbonaceous oxidation which occurs within the BOD bottle can be loosely described by two equations. The first outlines the oxidation of carbonaceous material into smaller molecules to provide energy which is then used in the synthesis of new cell material (reaction 1.2). The second describes the resynthesis of these small molecules into the cell components required by the growing microorganisms, utilising the energy generated in the first reaction (1.3). These two processes of degradation and resynthesis are known respectively as catabolism and anabolism and will be covered in more detail in Chapter 5.

1. *Catabolism*

$$CHON + O_2 \longrightarrow CO_2 + H_2O + \text{Small molecules} + \text{Energy} \tag{1.2}$$

Organic
material

2. *Anabolism*

$$\text{Small molecules} + O_2 + \text{Energy} \longrightarrow C_5H_7O_2N + H_2O \tag{1.3}$$

Towards the end of the incubation period when the amount of biodegradable organic material in the bottle is very low (Figure 1.2), many of the microorganisms are experiencing starvation conditions and are forced to oxidise their own cellular carbonaceous material in order to provide the energy necessary for their continued viability.

This is known as endogenous metabolism and may be expressed as

$$C_5H_7O_2N + O_2 \longrightarrow CO_2 + NH_3 + H_2O + Energy \qquad (1.4)$$

This reaction is responsible for the faint residual oxygen uptake which may be observed for several months. In addition, compounds which prove more resistant to oxidation will continue being oxidised at a slower rate for a much longer period of time, until after approximately 28 days, very little oxygen uptake is occurring.

The cumulative oxygen demand over this 28-day period is termed the ultimate BOD or BOD_u, and it is reasoned that as all biological oxidation has ceased after this point, BOD_u represents the biodegradable organic fraction of a wastewater. Any organic material remaining in the waste is resistant to biological oxidation and may thus be safely discharged to a watercourse, without the fear of it exerting any further oxygen demand.

It is frequently observed that for wastes which contain a significant fraction of ammonia (which includes all domestic wastes), there is a sharp increase in the cumulative oxygen uptake between 5 and 12 days. This results from the activity of an important group of bacteria known as the nitrifiers which have the ability to oxidise ammonia to nitrite and nitrate, according to the reaction scheme:

$$NH_4^+ + 2O_2 \longrightarrow NO_3^- + H_2O + 2H^+ \qquad (1.5)$$

The oxygen requirements for this reaction are known as the nitrogenous oxygen demand (NOD). Obviously, as ammonia is an inorganic compound, this reaction will interfere with the calculation of carbonaceous content. Fortunately, nitrifying bacteria are susceptible to inhibition by a wide variety of inhibitors which do not affect the oxidising ability of non-nitrifying bacteria. It is usual, therefore, to perform the BOD test in the presence of a nitrifying inhibitor such as 1-allyl-2-thiourea (ATU).

Dependent upon the initial rate of carbonaceous oxidation, the decline in the rapid uptake of oxygen may occur over a wide time period (Figure 1.3). Consequently two samples of identical carbonaceous content but with different rates of oxidation, would show different values for their BOD depending upon which day the test was terminated. This discrepancy was well known to the UK Royal Commission which was formed in 1898 to investigate methods for the treatment and disposal of sewage, and they carried out many experiments in an attempt to resolve this problem. Their results were published in 1912 in their eighth report entitled *Standards and Tests for Sewage and Sewage Effluents Discharging into Rivers and Streams*. In this report they recommended that a period of 5 days' incubation be employed, as it was found to give a smaller experimental error than longer incubation periods. In addition the Royal Commission also recommended that the incubation temperature be 65 °F as this was the maximum river temperature in the UK in the hottest month. This

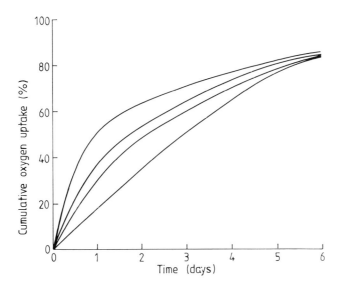

Figure 1.3 The effects of initial oxidation rates on the shape of the cumulative oxygen uptake curve.

figure converts to 18.3 °C which has been rounded up for convenience to 20 °C. The test has remained little changed since 1912 and is known universally as the BOD_5^{20} test.

In order to determine the BOD of a given wastewater, it is not necessary to measure the oxygen uptake on each of the 5 days, but merely to take two readings, one at the start and one at the end of the incubation period. It is important, however, to ensure that at the end of the incubation period there is still residual dissolved oxygen present in the bottles. Obviously, once residual dissolved oxygen has been completely depleted, no more oxidation can occur even though unoxidised waste may remain. Complete depletion of dissolved oxygen indicates that the dilution of the original sample was insufficient and that the test must be repeated using a larger sample dilution.

Kinetics of the BOD test.   Although the BOD test comprises a series of complex biochemical reactions carried out by microorganisms, it has still proved possible to establish empirical mathematical models which adequately describe the observed experimental data. The rate of biochemical oxidation of organic matter has been shown to be proportional to the concentration of organic material which remains unoxidised; thus it may be represented by a first-order reaction of the form:

$$\frac{dL}{dL} = -k_1 L \qquad (1.6)$$

where L is the BOD remaining at any given time (mg O_2/l) and k_1 the first-order BOD removal constant (d^{-1}).

If the amount of BOD exerted at zero-time (t_0) is L_0, and after a given time t this is oxidised to an amount L, then equation (1.6) may be integrated between the limits t and t_0 to yield:

$$\ln \frac{L}{L_0} = -k_1 t \tag{1.7}$$

which may be rearranged giving

$$L = L_0 \exp(-k_1 t) \tag{1.8}$$

The amount of BOD which has been exerted at any time t, can be defined by the term Y as

$$Y = L_0 - L \tag{1.9}$$

and if the term L in equation (1.9) is now replaced with $L_0 \exp(-k_1 t)$ from equation (1.8), it may be rearranged to give

$$Y = L_0[1 - \exp(-k_1 t)] \tag{1.10}$$

The above relationships are valid only for data gathered at constant temperatures, since the rate of BOD removal is very temperature dependent. Although the standard temperature for the BOD test is 20 °C, it may be fact be performed at any temperature providing that this is maintained constant and quoted along with the results. Indeed, for many hot climates, BOD tests performed at an elevated temperature of say 25 or 28 °C would provide more meaningful information. For the purpose of comparison with data obtained at 20 °C, a simple temperature function may be used to correct the first-order rate constant. This takes the form:

$$k_t = k_{20}\theta^{(t-20)} \tag{1.11}$$

Many different values have been quoted for θ, but a figure of 1.047 is generally used for temperatures above 20 °C and a figure of 1.135 for temperatures below this.

Armed with equations (1.10) and (1.11), it is possible to calculate the BOD exerted by a sample for any given incubation period and for any given incubation temperature, provided that the first-order rate constant is known, together with a BOD value for at least one incubation period.

Chemical oxygen demand

This test utilises potassium dichromate in boiling concentrated sulphuric acid (150 °C), in the presence of a silver catalyst, to act as a strong oxidising agent. Under these conditions most of the carbon in the sample is oxidised to CO_2 and the hydrogen present is oxidised to water. At the same time the dichromate is reduced to trivalent chromium. This reaction is illustrated using potassium phthalate as an example of an organic compound:

$$2KC_8H_5O_4 + 10K_2Cr_2O_7 + 41H_2SO_4 \longrightarrow 16CO_2 + 46H_2O$$
$$+ 10Cr_2(SO_4)_3 + 11K_2SO_4 \qquad (1.12)$$

In the above reaction each molecule of potassium dichromate has the same oxidising power as 1.5 molecules of oxygen. Thus 2 molecules of potassium phthalate consume 15 molecules of oxygen. Consequently the more organic material present in the sample, then the more dichromate will be reduced to chromium. The COD of the sample is determined by titrating the remaining dichromate with ferrous sulphate. The ferrous ions in the titrant react with chromic ions as follows:

$$3Fe^{2+} + Cr^{6+} \longrightarrow 3Fe^{3+} + Cr^{3+} \qquad (1.13)$$

A indicator known as Ferroin (1,10-phenanthroline) is used to indicate the end-point of this reaction. This compound forms an intense red colour with ferrous ions (Fe^{2+}) but no colour with ferric ions (Fe^{3+}). Thus when the added ferrous ions have finished reacting with the Cr^{6+} ions, the colour of the solution changes sharply from greenish blue to a reddish brown.

Although the COD test provides a very strong oxidising environment, certain aromatic compounds (e.g. pyridine, benzene, toluene) are not oxidised, and neither is ammonia.

Permanganate value (PV)

This principles of this test are very similar to that of the COD test in that the organic material in the waste is oxidised chemically, and the amount of oxidising agent remaining after the reaction is determined by titration. In this case the conditions of oxidisation are much milder; the oxidising agent is potassium permanganate in the presence of dilute sulphuric acid, and the sample is boiled for 30 min. Upon oxidation, the bright pink permanganate solution is reduced to a colourless liquid, and this offers a convenient visual guide as to whether the dilution of the wastewater sample is large enough. If not the permanganate will become colourless before the end of the oxidation period and a greater dilution is required. The remaining permanganate is titrated with ammonium oxalate.

Because of the mild oxidising conditions employed in this test, many organic compounds are either only partially oxidised or not oxidised at all. In addition certain inorganic compounds (sulphide, thiosulphide and thiocyanate) are oxidised by permanganate, whereas ammonia is not.

Total oxygen demand

This is a rapid automated test which oxidises a waste in the presence of a catalyst at 900 °C in a stream of air. Under these harsh conditions all the carbon in the sample will be oxidised to CO_2 and the oxygen demand is calculated from the difference in oxygen content of the air before and after oxidation. The resulting value for TOD is all embracing as it includes oxygen used in the oxidation of organic and inorganic substances. In addition, commonly occurring non-degradable compounds such as lignin and plastics will also yield an oxygen demand.

Comparison of techniques for the evaluation of organic material

The four techniques presented above for the evaluation of the pollution potential of a wastewater each offer certain advantages and disadvantages. The choice of one of these as a universally acceptable assay has been the subject of controversy for decades. Indeed as long ago as 1914 an American Public Health Association (APHA) committee was established to investigate oxygen tests, and it was noted of one of the more distinguished participants: 'Professor Phelps finds the work not very satisfactory and it is his intention to develop some method which will take us entirely away from oxygen determination.' Unfortunately, however, such an ambition has not yet been realised, and despite a vast amount of research on this subject we are still left with oxygen determination.

Of the four tests available for determining oxygen demand, the BOD finds the most favour and is the most widely used world-wide. In part this stems from its antiquity, as engineers have now developed a 'feel' for the BOD test and understand the significance of a particular BOD value in a given context. In addition many people also claim that, as the BOD test is a bioassay, it stimulates the biological events occurring in the receiving watercourse and therefore provides an absolute index of the pollution potential of a wastewater. Unfortunately this is in fact far from the truth as the biochemical reactions which occur in the BOD bottle are very dissimilar from those occurring in a receiving water. Indeed it has been suggested that the BOD test is more a measure of the metabolic activity of the microorganisms present in the seed under a given set of conditions, than it is a measure of the strength of a sewage. In addition the incubation period of 5 days means that the test can only provide historical data and can thus play no part in pollution prevention. Despite the numerous disadvantages of this test it has a very strong following and is likely to be employed for many years to come.

By comparison, the COD test is comparatively new and, while it is also routinely determined at sewage treatment works, it is rarely used in effluent discharge control, but primarily in assessing the strength of trade effluents. Because the COD test is a simple chemical assay, of which the principle reactions are well understood, it is easy to point out its drawbacks and limitations. This is probably one of the reasons for its unpopularity. Conversely, since the BOD test is a bioassay, it is very poorly understood and it is difficult, therefore, to appreciate exactly what its limitations are.

As a result of the strong oxidising conditions employed in the COD test, values obtained are considerably higher than those in the BOD test and a relationship of COD:BOD_5 of 2:1 is routinely observed for municipal settled sewage. However, this ratio changes as the waste passes through the plant (usually increasing), as the biodegradable fraction of the waste decreases whereas the non-biodegradable fraction remains unchanged. The fact that the COD test will return an oxygen demand for organic material which is non-biodegradable, makes it appear unsuitable in many people's eyes as an indicator of pollution potential. However, this problem is easily circumvented by the use of ΔCOD, where the COD of a waste is determined before and after treatment. The difference is the total amount of oxygen required to treat the wastewater. As the amount of oxygen utilised by microorganisms metabolising a waste is related to its organic content,

Test	Advantages	Disadvantages
BOD	Simple, popular. Used in the majority of design equations. Familiar to most engineers. Produces information on both carbonaceous and nitrogenous oxygen demand	Long period of incubation. Reproducibility poor ($\pm 15\%$). Susceptible to inhibition by many industrial wastes
PV	Simple inexpensive apparatus required. Rapid (data available within 40 min). Ideal for field testing. Good reproducibility ($\pm 6\%$).	Many organic compounds are not oxidised by the mild conditions. Certain inorganic compounds may contribute a high oxygen demand
COD	Simple inexpensive apparatus required. Comparatively rapid (data available within 3 h). ΔCOD gives an accurate indication of the fraction of the waste amenable to biodegradation. Good reproducibility (± 5–10%)	Does not oxidise ammonia. Many non-biodegradable organic compounds exert an oxygen demand. Interference from high concentrations of chloride ions
TOC	Very rapid (data available in minutes). May be readily automated. Reproducibility excellent (± 3–6%)	Expensive apparatus and skilled technician needed. Little comparative data available

Table 1.1 Relative merits of the tests available for the determination of oxygen demand.

it can be seen that ΔCOD is a measure of the amount of carbon source in the sample that was utilised as a food source by the microorganism. In other words, it is an absolute measure of the fraction of waste amenable to biodegradation by a particular treatment process. In addition it can give an accurate indication of the fraction of the waste which will remain undegraded. In view of its simplicity and rapidity the COD test is the most suitable assay for the determination of the strength of both raw and treated wastewaters.

The remaining two tests are both similar to COD and find application in certain specialised environments. Because the PV test is very rapid, simple and requires a minimum of laboratory equipment, it is ideal both for the field testing of water and also for laboratories in developing countries where little sophisticated equipment is available. In contrast the TOD test requires expensive apparatus and well-trained operators. However, because it is such a rapid assay (as short as 5 minutes), it can be automated and used for on-line process control. This is particularly useful for factories or sewage works which are discharging into an environmentally sensitive watercourse, or where strict regulations are imposed to limit the strength of wastewater discharges. The advantages and disadvantages of each of these tests are summarised in Table 1.1.

Theoretical oxygen demand

Linear relationships are generally found to exist between each of the above four assays, with the relative strength being in the order:

$$PV < BOD < COD < TOD$$

Theoretically, however, for a waste which is completely oxidised in all of the above assays, the value for the oxygen demand should be identical in each case. This value is known as the theoretical oxygen demand (ThOD) and is easily calculated if the appropriate chemical formulae are known. As the waste is completely oxidised, then any carbon molecules will be converted to CO_2, hydrogen to H_2O, nitrogen to NO_3, sulphur to SO_3, etc. Thus for a common waste product—urea:

$$CH_4N_2O \longrightarrow CO_2 + H_2O + NO_3 \qquad (1.14)$$

This equation now requires balancing in terms of oxygen molecules, thus:

$$CH_4N_2O + 9/2O_2 \longrightarrow CO_2 + 2H_2O + 2NO_3 \qquad (1.15)$$

It is apparent from the above equation that every molecule of urea requires 9/2 molecules of oxygen for its complete oxidation. This may be expressed in terms of weight by consideration of the molecular weight of the reactants. Hence $(12 + 1 \times 4 + 2 \times 14 + 16)$ gram-equivalents of urea require $(9/2 \times 16)$ gram-equivalents of oxygen for complete oxidation. Therefore:

$$1 \text{ g urea requires } 1 \times \frac{72}{60} = 1200 \text{ mg O}_2/\text{l}$$

Thus the theoretical oxygen demand of a solution containing 1 g/l of urea is 1200 mg O_2/l.

1.3 CHARACTERISATION OF INORGANIC COMPONENTS

Unlike organic material, there is no simple assay available which is equivalent to the oxygen demand tests, and allows a gross determination of the pollution potential of the inorganic material present in a sample. Fortunately, however, the number of inorganic compounds which pose the threat of serious pollution are limited, and it is quite feasible to perform simple assays to detect those individual compounds which are most likely to prove troublesome. The most important of these compounds will be covered on an individual basis.

Nitrogen and phosphorus

Nitrogen and phosphorus in a watercourse originate from many sources including: artificial fertilisers applied to farmland, farm animal waste, many manufacturing processes and in particular effluents from sewage treatment works. The nitrogen in sewage effluent arises primarily from metabolic interconversions of excreta-derived compounds, whereas 50% or more of the phosphorus arises from synthetic detergents. The principal forms in which they occur in watercourses are: NH_4^+ (ammonium); NO_2^- (nitrite); NO_3^- (nitrate); PO_4^{3-} (orthophosphate).

Together these two elements are known as nutrients and their removal is known as nutrient stripping. Their presence in a watercourse causes pollution by two means. As shown earlier, a group of microorganisms known as the nitrifying bacteria, are capable of oxidising the ammonium ion to nitrate, and in doing so will utilise oxygen dissolved in the watercourse with 1 mg of ammonia requiring up to 4.5 mg O_2 for its complete oxidation. As the saturated oxygen concentration of water is only 9.0 mg O_2/l at 20 °C, it requires only a small concentration of ammonium in a discharge under the appropriate conditions to completely deoxygenate a body of water. Such conditions are comparatively rare in temperatre climates as the nitrifying bacteria have an optimum temperature of 20 °C and suffer inhibition below this; however, in hot climates, very high rates of nitrification are common.

In addition to its propensity for oxidation by nitrifying bacteria, ammonia in solution is highly toxic in its own right, proving lethal to most species of fish at low concentrations. Ammonia is highly soluble in water, establishing an equilibrium of un-ionised ammonia (NH_3), ionised ammonium (NH_4^+) and hydroxyl ions (OH^-), according to the reaction scheme:

$$NH_3 + H_2O \longleftrightarrow NH_3.H_2O \longleftrightarrow NH_4^+ + OH^- \qquad (1.16)$$

The toxicity of ammonia in solution is directly attributable to the NH_3 species, and a limit of 0.025 mg NH_3/l has been established for European fisheries by the European Inland Fisheries Advisory Committee (EIFAC), as the maximum concentration of un-ionised ammonia which can be safely tolerated by fish over an extended period. A slightly lower figure of 0.02 mg/l un-ionised ammonia was established by the Environmental Protection Agency (EPA) to cover North America. Because of the equilibrium which exists between NH_3, NH_4^+ and OH^-, the toxicity of ammonia is pH dependent, increasing with increased pH values. In addition, however, the concentration of NH_3 is also found to increase with increasing temperature. As a result of this the standards recommended by the EIFAC and the EPA apply to temperatures above 5 °C and pH values below 8.5. The combined effects of temperature and pH on the availability of un-ionised ammonia are illustrated in Figure 1.4.

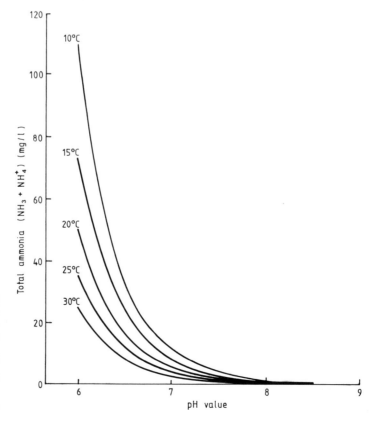

Figure 1.4 The concentration of total ammonia required to produce an un-ionised ammonia concentration of 0.02 mg/l, for a range of temperatures and pH values.

Perhaps the most widespread example of pollution through nitrogen and phosphorus discharges occurs through their ability to promote the growth of algae. In an aquatic ecosystem the size of the algal population appears to be limited by the concentration of nitrogen and phosphorus. As algae possess chlorophyll and are thus capable of photosynthesis, they do not need to oxidise carbon to obtain their energy and can live in aquatic environments where organic compounds are absent or at low concentrations. They do, however, require a plentiful supply of nutrients in order to synthesise their cell components, and in the presence of an excess of nitrogen and phosphorus vast blooms of algae are common. In such a situation the watercourse is said to be eutrophic. Apart from the unsightly nature of these algal blooms, they can also prove ecologically harmful to the other inhabitants of the water. During the day when photosynthesis occurs, algae produce large quantities of oxygen which help to aerate the watercourse; in addition, however, they also utilise dissolved carbonates and bicarbonates as their source of cell carbon. This results in an increased water pH, with values as high as 10.5 being observed. Conversely, at night the reverse reaction occurs when the algae respire, taking up oxygen and giving off CO_2. As a consequence the water is subjected to widely fluctuating pH values, and in addition if the algal concentration is sufficiently high, the water may become completely anaerobic at night. This condition is extremely common world-wide, particularly where a landlocked lake receives either raw or treated sewage. Typical eutrophic waters include Lake Erie in North America, Lake Como in Italy, Lake Kasimigaura in Japan and Lake Mariout in Egypt. A limiting figure for phosphorus of 100 μg/l in a flowing stream and 50 μg/l in a stationary body of water, is often quoted to prevent biological nuisance.

Another feature of eutrophic waters is the lack of species diversity. Whereas in an oligotrophic water (one which has a very low nutrient concentration), the algae are sparse in the number of individuals, but remarkably diverse in numbers of species. Under eutrophic conditions the algae are abundant in number, but certain species come to be strongly dominant. This observation allows a rapid microscopic assessment of the nutrient status of a body of water by examination of its algal diversity.

Where the water receiving the discharge is used for abstraction purposes, algae present a different problem owing to the difficulties associated with their removal. A typical eutrophic reservoir in London of $27 \times 10^6 \, m^3$ capacity, may well contain up to 110 dry tonnes of algae. As the retention time of the reservoir is typically 10 days, it is necessary to remove 11 dry tonnes of algae each day from the abstracted water. By contrast a typical oligotrophic lake of the same size contains as little as 3 dry tonnes of algae.

A further problem of nitrogen pollution arises where water receiving effluent discharges is used for abstraction, and the nitrogen is present in the form of nitrate. Nitrate pollution of drinking-water poses a direct risk to human infants as a result of the microbial interconversions which occur in an infant's intestinal tract. Below the age of 6 months, infants do not possess a fully developed digestive system, and as a consequence of this the pH of their stomach is in the region of 4.0, as compared to adults who have stomach pH values of 2.0. This results in them having a different stomach microbial flora to those of adults and included in this flora is a bacterium which is capable of reducing nitrate to nitrite. The nitrite is then reoxidised to nitrate using haemoglobin as an oxidising agent. This is achieved according to the reaction:

$$NO_2^- + \text{Haemoglobin} \longrightarrow NO_3^- + \text{Methaemoglobin} \quad (1.17)$$

The reduced form of haemoglobin, methaemoglobin, lacks the ability to bind with oxygen and this results in the death of the infant due to oxygen starvation. This condition is known variously as: methaemoglobinaemia, infantile cyanosis or the blue-baby syndrome. The dangers of infantile cyanosis have led the World Health Organisation to declare a maximum permissible figure of 100 mg/l nitrate in drinking-water, whereas the European Economic Community 1985 Directive on drinking-water quality limits the maximum admissible concentration of nitrate to 50 mg/l. Sewage works effluents routinely contain 10–40 mg NH_4^+/l and 5–30 mg NO_3^-/l.

Toxic pollutants

In addition to the potentially toxic side-effects which result from the oxygen demand of organic discharges, many compounds prove toxic directly by interfering with the metabolism of an organism. Many thousands of toxic compounds such as pesticides, insecticides and the heavy metals are currently used in agriculture and industry. A measure of their relative toxicity is assessed by means of a test which measures the concentration of toxicant which is lethal to 50% of the population under test over a given exposure period, usually 48–96 hours. It is written as LD_{50} (lethal dose) or LC (lethal concentration). The test gives a comparison of the absolute toxicity of a compound in a single concentrated dose, known as acute exposure. Fish are routinely utilised as test animals and it is thought that a concentration of 10% of the LD_{50} value may safely be discharged into the environment without adversely affecting aquatic life. Thus, the dilution which a wastewater discharge must receive in order to reduce its toxic waste concentration to a safe level is given by

$$V = \frac{(100 - LD_{50})}{LD_{50}} \cdot \frac{Q}{10} \quad (1.18)$$

where V is the volume of dilution water required, and Q the volume of wastewater.

Information on LD_{50} values is of limited applicability, however, since it gives no indication of the long-term effects of sub-lethal concentrations of toxicants (known as chronic exposure), on an ecosystem. This is particularly pertinent for the non-biodegradable compounds such as insecticides and pesticides which are absorbed by bacteria and thus enter the aquatic food chain. These compounds will remain in the environment for long periods of time, slowly being concentrated at each stage in the food chain until ultimately they prove fatal, generally to predators at the top of the food chain such as fish or birds (Figure 1.5). Such a phenomena occurred in the small coastal village of Minamata Bay, Japan. The local inhabitants of this village have a diet comprised largely of fish and shellfish, and in the early 1950s many of the population suffered symptoms of headaches, blurred vision and impaired speech. These mild symptoms were then followed in many cases by paralysis and often death. The symptoms resulted from damage to the central nervous system and brain of the villagers, and the cause was eventually traced to mercury poisoning. The mercury originated in the effluent from a local chemical plant which utilised mercuric chloride, and was discharged into Minamata Bay. Microorganisms found in the sediment of the bay were capable of converting this mercuric chloride into methyl mercury—a more toxic form of mercury owing to the ease with which it is absorbed and concentrated by organisms. Consequently

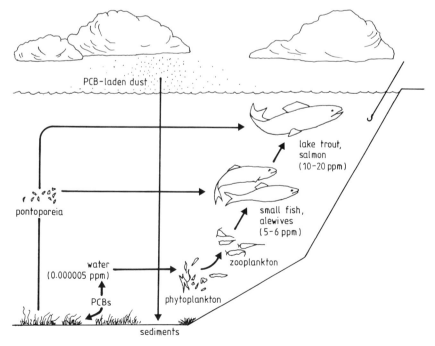

Figure 1.5 The concentration of PCBs within an aquatic food chain.

it moves rapidly through a food chain and accumulates at higher trophic levels. By 1975 3500 cases of mercury poison had been recorded at Minamata, of which 45 proved fatal.

As a direct result of the tragedy of Minamata, there is a growing awareness of the potential health risk resulting from the contamination of aquatic ecosystems by organic chemicals resistant to biodegradation. One particular group, known as chlorinated hydrocarbons, poses the most consistently unacceptable risk to the environment and human health. This group includes such compounds as DDT, aldrin, dieldrin, chloradane and also the polychlorinated biphenyls (PCBs). These industrial chemicals are widely used within the electronics industry and gain entry into the aquatic environment via treatment plant effluents, surface run-off from manufacturing sites and leachates from industrial refuse disposal sites. The accumulation of PCBs within an ecosystem is illustrated in Figure 1.5. Laboratory experiments have suggested that a steady diet containing 2.5 mg/d PCBs will cause serious health disorders.

The devasting consequences for aquatic life which result from the discharge of toxic pollutants into an aquatic ecosystem, has been further highlighted by the pollution of the River Rhine by the Swiss chemical firm Sandoz. As a result of a fire at a chemical warehouse near the city of Basle, Switzerland, in 1986, chemicals were washed by fire-hoses into the nearby Rhine. A total of 1300 tonnes of chemicals were burnt in the blaze including 900 tonnes of pesticide and 12 tonnes of organic compounds containing mercury. This resulted in a 200 km long slick of pollution containing pesticides and mercury levels up to $0.22\ \mu g/l$. The slick caused the death of 500 000 fish in the Rhine with the prospect of long-term pollution to follow as a result of pollutants trapped in the benthic layer. The disaster undid in hours the results of a 15-year campaign to clean up the Rhine which saw the number of fish species increase from three to fifteen, at a cost of some £14 billion.

1.4 CHARACTERISATION OF WASTEWATER SOLIDS CONTENT

The organic and inorganic components of wastewaters are present in both a soluble and insoluble (or particulate) form. The terminology used to denote the various forms of solids within a sample often proves quite confusing, with several terms being used to represent essentially the same form. Three distinct types of solids are recognised in wastewaters: suspended solids, dissolved solids and volatile suspended solids. The sum of the suspended and dissolved solids are known as the total solids. The boundary between suspended solids and dissolved solids is an arbitrary one and it is determined by filtration of the sample through a filter paper which has a pore size

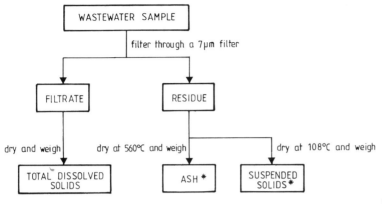

Figure 1.6 Determination of the solid fractions of a wastewater.

greater than $0.45\,\mu\mathrm{m}$. Material retained on the filter is termed insoluble solids although it is more precisely known as non-filterable residue. The liquid which passes through the filter contains the soluble solids (or filterable residue) and these are determined by weighing after evaporating away the water in a steam-bath. Both soluble and insoluble solids can comprise organic and inorganic material. These may be differentiated by heating the solids to a temperature of $560\,^{\circ}\mathrm{C}$, a process known as ignition. The loss on ignition is taken to represent the organic content of the sample, whereas the residual ash or fixed solids is a measure of the inorganic (mineral) content (Figure 1.6). Of particular importance from the point of view of wastewater treatment is a group of solids known as settleable solids. These are determined using an Imhoff cone and are represented by the volume of sludge in the bottom of the cone after settlement for 1 hour (Figure 1.7). Although the solids content of domestic wastewaters is reasonably homogeneous, it varies considerably for industrial wastewaters. The range of solids encountered in typical domestic and industrial wastewaters is shown in Table 1.2. Many attempts have been made to separate and define wastewater solids on the basis of their particle size and four distinct size fractions have been identified namely: settleable $> 100\,\mu\mathrm{m}$; supracolloidal $1-100\,\mu\mathrm{m}$; colloidal $1\,\mathrm{nm}-1\,\mu\mathrm{m}$; soluble $< 1\,\mathrm{nm}$.

Using such a system of classification it has been shown that the soluble solids in a wastewater comprise largely inorganic material, whereas suspended material is predominantly organic. Suspended solids can contribute up to 60% of the BOD of the wastewater and consequently they will undergo biodegradation and exert an oxygen demand, in a similar way to the dissolved solids. In addition, however, owing to their larger size and more rapid settling velocities, they may frequently settle quickly and form a sludge blanket near the point of the effluent discharge. The organic

Figure 1.7 The use of an Imhoff cone to determine the fraction of settleable solids in a wastewater.

Particle size					
10^{-8} 10^{-7} 10^{-6} 10^{-5} 10^{-4} 10^{-3} 10^{-2} 10^{-1} 1 10 10^2 (mm)					
10^{-5} 10^{-4} 10^{-3} 10^{-2} 10^{-1} 1 10 10^2 10^3 10^4 10^5 (μm)					

Particle form	True solutions	Colloidal suspensions	Suspended and floating solids		
Particle designation	Ionic	Macro-molecular	Micro-particle	Fine particle	Coarse particle

Table 1.2 Particle sizes of solids found in industrial and domestic wastewaters.

material deposited on the bed of the watercourse is now subjected to benthal biodegradation, which results in a depletion of the oxygen source at a faster rate than it is supplied from the watercourse. This material is therefore subjected to anaerobic breakdown which results in the production of methane and toxic hydrogen sulphide gases (see p. 272). The dissolved oxygen removed by this benthal breakdown of organic material, is one of the major reasons for the discrepancies in models of the oxygen sag curve (p. 35).

1.5 CHARACTERISATION OF THE MICROBIAL COMPONENTS OF WASTEWATER

A wide variety of microorganisms are found in wastewaters including: viruses, bacteria, fungi, protozoa and nematodes. These organisms occur in very large numbers, in particular the bacteria for which total counts in the range $1-38 \times 10^6 \, \text{ml}^{-1}$ are routinely recorded. It would be extremely time consuming, and provide little information of use for engineers, to identify all the species of microorganisms found in wastewaters. The main aim in the microbiological examination of a water, as far as the engineer is concerned, is to detect the presence of pathogenic microorganisms which would constitute a danger to human health through contact with the contaminated water. Between 20 and 30 different infections are associated with water usage, namely through water-borne, water-based, water-washed or water-related vectors. A comprehensive list of these pathogens is found in Table 1.3. All water-borne, and many of the water-based diseases, are dependent for their transmission upon the access of faeces from a contaminated individual (which therefore contain excreted pathogens), into a water supply utilised for drinking or bathing purposes. The health problems resulting from the presence of excreted pathogens in wastewaters are numerous and the scale of the problem large, particularly in developing countries. The number

Bacteria	Viruses	Intestinal parasites
Salmonella typhi	Enteroviruses	*Schistosoma* spp.
S. paratyphi	Poliovirus	*Ascaris lumbricoides*
Other spp.	Echovirus	*Trichuris trichuria*
Shigella spp.	Coxsackie viruses	*Taenia* spp.
Vibrio cholerae	New enteroviruses	*Diphyllobothrium latum*
Mycobacterium tuberculosis	Hepatitis type A	*Ankylostoma duodenale*
Leptospira icterohaemorrhagiae	Norwalk virus	*Necator americanus*
Campylobacter spp.	Rotavirus	*Entamoeba histolytica*
Listeria monocytogenes	Reovirus	*Giardia lamblia*
Candida albicans	Adenovirus	*Naegleria* spp.
Yersinia enterocolitica	Parvovirus	*Acantamoeba* spp.
Enteropathogenic *Escherichiacoli*		*Cryptosporidia*
Pseudomonas aeruginosa		
Klebsiella spp.		
Staphylococcus aureus		
Aeromonas hydrophila		
Mycobacterium paratuberculosis		
Erysipellothrix rhusopathiae		
Bacillus anthracis		
Clostridium spp.		
Yersinia pestis		
Brucella spp.		

Table 1.3 Pathogenic agents that may be present in sewage.

of people world-wide suffering from the water-related diseases listed in Table 1.3 is estimated by the World Health Organization at 1.25 billion. The majority of infections associated with the water cycle, however, are those which cause diarrhoea and this disease is estimated to kill 6 million children under 5 years of age each year in developing countries and is a contributory factor in the deaths of a further 18 million. There are as many as 65 000 cases of diarrhoea reported every hour world-wide. This disease follows a faecal–oral route whereby the infected individual suffers from rapid multiplication of the infecting agent in the gut, consequently faeces from these individuals contain elevated concentrations of pathogens which may then gain access to water. Any person subsequently consuming the contaminated water has a high risk of contracting the disease. A typical faecal–oral infection cycle is illustrated in Figure 1.8.

The pathogenic microorganism routinely encountered in wastewater have many origins. Human excreta from infected individuals, discharges from abattoirs and plants which process animal material and surface run-off from land which is grazed by animals, all provide a constant source of pathogens. It is apparent from Table 1.3 that direct examination of a water for the presence of each of these pathogens would be a lengthy process. In addition the isolation of these harmful

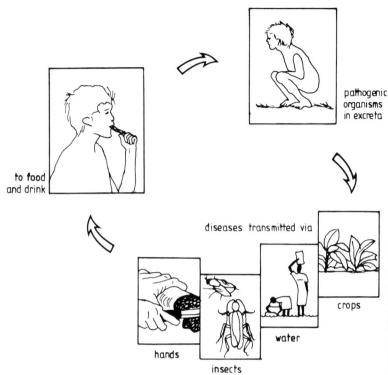

pathogenic
organisms
in excreta

to food
and drink

diseases transmitted via

crops

water

hands

insects

Figure 1.8 Faeco-oral cycle illustrating infection due to the transmission of a faecally borne pathogen. (From Winblad and Kilama, 1985. Reproduced by permission of Macmillans.)

organisms in a laboratory is not advisable owing to the potential risk to laboratory personnel. Consequently, other harmless organisms have been sought whose presence would reliably indicate the presence of pathogenic microorganisms. The search for such an organism has centred on the fact that pathogenic microorganisms responsible for transmission of water-borne diseases originate from contamination of a water source with faecal material from an infected person. Consequently, if it can be shown that the water source is contaminated with faecal material, then this will indicate that there is a *potential* health risk from the presence of excreted pathogens. Organisms which indicate the presence of faecal pollution are known as indicator bacteria and an ideal indicator bacteria should possess the following properties. They should:

1. Be present exclusively in faeces and must always be present when there is the likelihood of pathogens being present;

2. Occur in greater numbers than any pathogen in order to prevent their being diluted out in a receiving water;

3. Be easily isolated and enumerated;

4. Show slightly greater resistance to survival in a watercourse than any pathogen, including resistance to chlorination and ozonation;

5. Be non-pathogenic themselves to ensure that they do not pose a hazard to laboratory personnel.

Because of the diverse survival characteristics and nature of excreted pathogens, a single perfect indicator for all pathogens is impossible. Despite this problem many organisms have been proposed as indicators of faecal contamination, of which three given under the headings below, have found widespread usage.

1. Faecal coliform

The most common member of this group is the organism *Escherichia coli* which has been widely adopted as an indicator in Europe and the United States. It is defined as a gram-negative rod which will produce acid and gas from the fermentation of a lactose peptone medium within 24–28 hours, and also indole from the degradation of the amino-acid tryptophan. Many members of the coliform group are capable of performing these reactions at 37 °C and they are known as total coliforms. Only if the organism is capable of performing the reactions at 44 °C is it known as a faecal coliform. Faecal coliforms other than *E. coli* include certain *Citrobacter* and *Klebsiella* species. Whereas the faecal coliforms are exclusively faecal, other coliforms also occur naturally in unpolluted waters and soils as well as in faeces. It is undoubtedly the most widely used faecal indicator and its behaviour in temperate environments is well characterised. However, when used in tropical climates its behaviour may not be the same and results require interpretation with caution. For instance in the tropics certain bacteria of non-faecal origin fulfil the definitions of a faecal coliform.

2. Faecal streptococci

This group of organisms was proposed as indicators in 1964 and have been adopted in Europe and the USA in a complementary role with the faecal coliforms. They are gram-positive cocci with a diameter of $\sim 1\,\mu$m and occur in short chains of up to three cells. This group comprises several species of streptococci some of which are exclusively human in origin, some occur in Man and Animals, whereas others have been reported in both polluted and non-polluted aquatic environments. Since certain biotypes are of non-faecal origin, and these are not distinguishable by routine detection methods, then the usefulness of the faecal streptococci is obviously limited.

3. Clostridium perfringens

This organism is a gram-positive, anaerobic, spore-forming rod which is identified by its ability to reduce sulphites to sulphides and

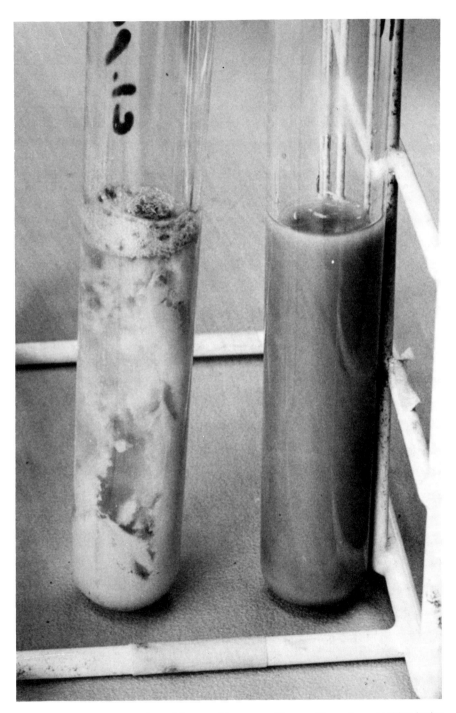

Figure 1.9 The growth of *Clostridium perfringens* on litmus milk medium resulting in the production of a stormy clot.

form a 'stormy clot' in litmus milk (Figure 1.9). This bacteria has a spore stage which is resistant to environmental stress, in particular desiccation and heat (remaining viable at 70 °C). Consequently it is very persistent in the environment and is used to indicate intermittent or remote pollution.

The presence or absence of the above organisms in a wastewater may lead an engineer to exercise his judgement in ways which have profound and far-reaching implications for human health. It is important, therefore, to understand exactly what the implications and limitations of laboratory results mean. The presence of indicator organisms does not mean that pathogens will necessarily be present, merely that faecal material is present and thus that there is a risk of the presence of pathogenic organisms. In addition, since all the faecal indicator bacteria described above occur in both human and animal faeces, it is not possible to delineate the source of the contamination. In addition, absence of indicators does not always mean the absence of pathogens, particularly if there is a danger of protozoal or helminthic infections (as the cysts and eggs from these organisms can survive a long time in the environment). Faecal indicator organisms find diverse applications in that they are expected to yield information on (supposedly) uncontaminated potable water, as well as on heavily contaminated wastewater. Because of this diversity in application the interpretation of faecal coliform counts on wastewaters from diverse sources will be considered on an individual basis in future chapters, as and when they arise.

1.6 COMPOSITION OF TYPICAL WASTEWATERS

Sources of pollution

Watercourses receive pollution from many different sources which vary both in strength and volume. Identification of the major sources of these pollutants is important, as they can be controlled most easily at source. Typical sites include:

1. Discharge of either raw or treated sewage from towns and villages;

2. Discharges from manufacturing or industrial plants;

3. Run-off from agricultural land;

4. Leachates from solid waste disposal sites.

The chemical and biological composition of the wastewaters from each of these sites will vary dramatically and determination of their composition is an essential task for an engineer. Where effluent treatment works have yet to be installed, detailed analysis provides information as to what type of treatment will prove the most suitable. Where treatment works have already been provided, analysis provides essential information which allows the works to be operated economically and efficiently. In addition, regular monitoring ensures that plant performance is maintained and that the treated wastewater does not pose an unnecessary threat to receiving water quality.

In addition to these continuous discharges, there is also the danger of accidental discharges of highly toxic material such as acids, cyanides or oil. The effects of these are often immediate and dramatic, with the watercourse suffering severe long-term damage. Obviously such events are difficult to control without foresight, and it is important in these situations to ensure adequate legislation which places the burden of rectifying the damage with the polluter.

Domestic and industrial effluents

Freshly discharged domestic sewage is a grey, turbid liquid with a characteristic, but (surprisingly) not unpleasant smell. However, if it is not kept fully aerated, it rapidly undergoes anaerobic biological degradation and this results in the production of such noxious compounds as hydrogen sulphide, amines and mercaptans which give it an extremely unpleasant smell.

The main source of both chemical and biological pollution in domestic sewage results from human excreta with a smaller contribution from wastewater resulting from laundry, food preparation and bathing (which are known collective as sullage). The composition of typical domestic sewages is illustrated in Table 1.4.

Determinand	Manchester (UK)	Mafraq (Abu Dhabi town)	Campina Grande (NE Brazil)	Amman (Jordan)	Nairobi (Kenya)
BOD (mg O_2/l)	240	228	240	770	520
COD (mg O_2/l)	520	600	570	1830	1120
PV (mg O_2/l)	—	75	—	—	—
Suspended solids (mg/l)	210	198	392	900	520
Ammonia (as N) (mg/l)	22	35.2	38	100	33
pH	7.4	7.6	7.8	—	7.0
Temperature (°C)	14	—	26	22	24

Table 1.4 Composition of typical domestic sewage.

By comparison trade wastes comprise spent liquor from manufacturing processes which are very strong, together with weaker water resulting from other processes such as washing, rinsing and condensing. Their final composition will obviously be dependent upon the nature of the manufacturing industries (Table 1.5). The total pollution

Determinand	Pharmaceuticals (India)	Textiles (India)	Beet-sugar waste (USA)	Coke-oven liquor (UK)
BOD$_5$	15 250	2000	930	1200
COD	28 540	5000	1601	3900
Suspended solids	5400	4000	1015	950
Ammonia (as N)	—	—	6.3	450
pH	9.3	12.0	7.1	5.5
Total N	5166	—	16.4	490

Table 1.5 Composition of effluent from typical industrial premises.

load resulting from industrial wastewaters is usually at least as great as that of domestic sewage. The final composition of a wastewater arriving at a works will obviously be dependent on the fraction of industrial waste which it contains. Because of the higher strength of the industrial waste, the operating regime of the factories which provide the discharges will obviously play a major part in determining the diurnal load variations observed with sewage influents, particularly when factories discharge on a batch basis.

In domestic sewage there is a tendency for the peak ammonia concentration to coincide with or occur just before the peak flow. However, in industrial wastewaters it is likely that the majority of the ammonia results from the effluents of a limited number of industrial discharges such as fertiliser manufacturing and chemical plants. In these cases the distribution of the ammonia load will be a function of the operating regime practised at the individual factories.

A factor which significantly affects the wastewater strength is the retention time in the sewers. Compositional changes will be reduced where there is a long sewer-rentention time. In addition, daily variations in strength will not be as great in sewerage systems which separate storm waters from domestic and industrial wastes as compared to combined storm and wastewater sewers.

Leachates from landfill sites

Unlike domestic sewage, which has characteristics within a fairly narrow range, leachates vary greatly in composition. The water present in a landfill site results from direct precipitation and groundwater movement, consequently the rate of leaching from a landfill site will be dependent upon the rate of infiltration from these two sources. In addition, seasonal and hydrologic factors also affect the composition of the ensuing leachate. Leaching will not occur until the site is fully saturated and this process may take several years. Inside the landfill site anaerobic conditions are quickly established and a high degree of biological activity is occurring resulting in the degradation and solubilisation of the site contents, thus the composition is changing constantly with the age of the site.

Determinand	Leeds (UK)	Bucks County (USA)	New York (USA)
BOD$_5$ (mg O$_2$/l)	6 000	12 500	10 040
COD (mg O$_2$/l)	12 000	18 500	7 500
Suspended solids (mg/1)	600	686	900
Ammonia (as N) (mg/l)	—	70	150
pH	6.2	6.7	4.3
Total N (mg/l)	—	748	350
Alkalinity (CaCO$_3$, mg/l)	—	5 500	—

Table 1.6 Composition of leachates from typical landfill sites.

Consequently, leachates generally have a high organic strength, low pH and contain very high heavy metal concentrations (Table 1.6). The wide variations observed in the composition of leachates means that a treatment process which has proved applicable at one site may not be directly transferable to other locations. In addition, due to the changes in leachate composition which occur as the landfill ages, the treatment process initially selected may well prove inappropriate after several years of operation.

1.7 EFFECTS OF ORGANIC DISCHARGES ON RECEIVING WATER QUALITY

Introduction

It has been shown that a wastewater represents an extremely complex mixture of organic and inorganic material, as well as supporting a complex microbial ecosystem. Before suitable standards can be imposed to limit discharges of such wastewaters, it is necessary to understand how an effluent reacts with its environment on discharge. This is important not only to ensure that the standard is strict enough to protect the environment and community served by the receiving watercourse, but also to ensure that it is not excessively conservative, thereby placing an unnecessary financial burden on the individuals who support the treatment facilities. Wastewater discharge standards should adequately reflect the potential dangers which a discharge poses, and as these dangers will vary greatly from environment to environment the concept of a universally applicable standard is unsound. The rest of this chapter, therefore, will explore the effects of wastewater discharges on natural ecosystems and with this information it should be possible to suggest rational effluent standards for any given environment.

Watercourses are constantly receiving discharges from numerous sources such as precipitation and surface water run-off, as well as the more obvious sources such as factories and sewage treatment works. Inevitably many of these point sources will have picked up a certain degree of contamination (measured in terms of oxygen demand) before reaching the watercourse. The way in which a given discharge will alter the concentration of organic material in a receiving watercourse will depend upon the strength and flow of both the discharge and the receiving water. The way in which an organic discharge affects the concentration of BOD in a watercourse below the discharge point may be calculated quite simply by deriving a mass-balance equation immediately below the point of mixing. Although it is usual in this circumstance to use BOD as a measure of pollution potential, the same principle may be applied to many other parameters, for instance COD, ammonia, nitrate or even the numbers of bacteria present in a water. Thus if a sewage treatment works discharges an effluent with a flow of Q_d and a BOD of L_d into a river which has a flow of Q_r and a BOD of L_r above the discharge point, as shown below,

Discharge

(Q_d, L_d)

↓

Q_r, L_r	River	$(Q_r + Q_d)\, L$

then the BOD of the river below the discharge point (L) is found by constructing a mass-balance equation around the point of discharge:

$$Q_r L_r + Q_d L_d = (Q_r + Q_d)L \qquad (1.19)$$

This approach was used by the UK Royal Commission in 1912 to arrive at an effluent discharge standard suitable for UK sewage treatment works. They assumed that a clean river should have a BOD of no more than 2 mg/l (L_r) and that in order to preserve the quality of a river, it should not rise above 4 mg/l (L). In addition they also decreed that at least an eightfold dilution of river water should be available ($Q_r/Q_d = 8$). Rearranging equation (1.19) into the appropriate form gives:

$$L_d = L[Q_r/Q_d) + 1] - L_r(Q_r/Q_d) \qquad (1.20)$$

Substituting the Royal Commission recommendations into this equation yields:

$$L_d = 4(8 + 1) - 2 \times 8 = 20\,\text{mg/l}$$

This figure was recommended as the maximum permissible effluent BOD for a sewage treatment works. In addition the Royal Commission knew that the majority of existing sewage treatment works were capable of producing an effluent containing 30 mg/l suspended solids and this figure was recommended as the maximum effluent solids concentration. Their 20/30 standard is known as the Royal Commission standard.

Monte Carlo simulation

The use of mass balances to estimate river quality downstream of an effluent discharge, assumes that if annual mean values or percentile values of flow and quality are substituted into equation (1.20), then mean or percentile values of downstream quality can be predicted. In reality this is rarely the case since the downstream river quality depends upon interactions between upstream flows and qualities. This criticism has been overcome by means of a computer simulation known as the Monte Carlo technique, which relates mean and percentile values of the downstream river quality to percentile values of the discharge quality. Assuming that the distributions of Q_r, L_r, Q_d and L_d are known, then a random value is taken from each distribution and used to calculate L. This random selection is repeated many times until enough values have been obtained to define the distribution of L. A mean and 95 percentile value for L can thus be obtained from the resulting distribution. In a similar way, if the distribution of L is known, then a distribution can be obtained for the effluent discharge quality, and a mean and 95 percentile discharge standard imposed.

Self-purification

All watercourses are capable of absoring a certain amount of pollution without suffering adverse effects. This is a result of the natural biological cycle which establishes a food chain of dependent communities capable of degrading organic material (Figure 1.10). This ability of a water to absorb and degrade organic pollution is known as its 'self-purification capacity'. Only when this capacity is exceeded will a water start exhibiting signs of pollution.

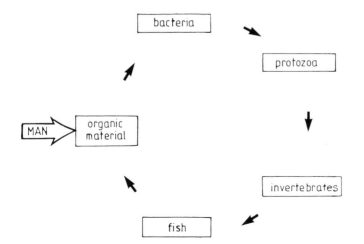

Figure 1.10 A simple example of an aquatic food chain.

The oxygen sag curve

Deaeration

On entering a watercourse, organic material present in the discharge will act as a food supply for the indigenous microorganisms and will be oxidised in a series of oxidation reactions similar to equation (1.1). The necessary oxygen for this reaction is obtained from the oxygen dissolved in the water, and the water is consequently deaerated. This deaeration is expressed in terms of a dissolved oxygen deficit, which is the difference between the maximum concentration of oxygen which the water can hold and the actual measured dissolved oxygen concentration. In other words it is the amount of oxygen required to bring the water up to its maximum oxygen saturation. The rate at which deaeration occurs is directly proportional to the BOD of the watercourse, i.e. it is first order with respect to BOD and can be expressed as

$$\frac{dD}{dt} = -k_1 L \tag{1.21}$$

where dD/dt is the rate of change in the dissolved oxygen deficit, k_1 the first-order BOD rate constant and L the concentration of BOD remaining in the water.

In order to use the concentration of BOD present in the water at the point of discharge (as determined from the mass-balance equation 1.20) it is necessary to substitute L in equation (1.21) with L_0 as given by equation (1.8). Thus:

$$\frac{dD}{dt} = -k_1 L_0 \exp(-k_1 t) \tag{1.22}$$

Reaeration

In addition to the oxygen depletion resulting from the microbial oxidation of organic material, waters are simultaneously being reaerated as a result of oxygen transfer between the atmosphere and the surface of the water in contact with the atmosphere. The rate at which this occurs is a function of the topographical features of the water such as depth, velocity, temperature and turbulence. The driving force for this oxygen transfer is the oxygen concentration gradient between the atmosphere and the body of water and consequently it is proportional to the oxygen deficit, i.e.

$$\frac{dD}{dt} = -k_2 D \tag{1.23}$$

where k_2 is the first-order reaeration constant.

The cumulative effect of reaeration and deaeration is to produce a characteristic curve known as an oxygen sag curve (Figure 1.11),

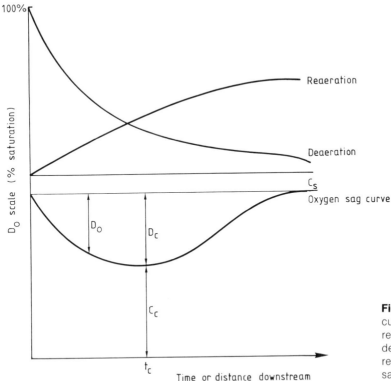

Figure 1.11 The cumulative effects of reaeration and deaeration which result in an oxygen sag within a river.

which shows the oxygen concentration at any point along a body of water below the point of waste discharge. The main points of interest to an engineer on this curve are:

1. The oxygen deficit at a given point downstream of the discharge;

2. The value of the critical oxygen deficit (i.e. the maximum oxygen deficit which the water will experience as a result of the waste discharge);

3. The time at which the critical oxygen deficit is reached (and thus the point on the watercourse where the maximum pollution will be exerted).

If this information is available, the strength of a waste discharge can be limited to ensure that the value of the critical oxygen deficit does not rise above a predetermined acceptable level. The objective of water-quality management is generally to maintain the dissolved oxygen concentration of a watercourse greater than this predetermined minimum concentration. In order to achieve this it is necessary to be able to predict the oxygen demand of a given discharge on the receiving water. Mathematical models are used to make such predictions and one of the simplest, known as the Streeter–Phelps model, is widely used to model the oxygen sag curve. In this model the oxygen content of a segment of water is determined as it moves downstream. This is determined by the construction of a mass-balance equation between two competing processes, namely oxygen uptake from the water which is used for carbonaceous BOD removal, and addition of oxygen from the atmosphere as a result of reaeration.

The combined effects of deoxygenation and reaeration upon the oxygen deficit can be expressed as

$$\frac{dD}{dt} = \text{Reaeration} - \text{Deaeration} \tag{1.24}$$

$$= k_1 L - k_2 D$$

substituting from equation (1.8):

$$\frac{dD}{dt} = k_1 L_0 \exp(-k_1 t) - k_2 D \tag{1.25}$$

Equation (1.25) is a first-order differential equation of the form:

$$\frac{dY}{dt} + PY = Q \tag{1.26}$$

where $Y = D$, $P = k_2$, $Q = k_1 L_0 \exp(-k_1 t)$.
It may thus be integrated using the integrating factor $\exp(-k_1 t)$, and for the boundary conditions $D = D_0$ when $t = 0$, the following equation is obtained:

$$D_t = \frac{K_1 L_0}{k_2 - k_1} [\exp(-k_1 t)] + D_0 \exp(-k_1 t) \tag{1.27}$$

where L_0 is the BOD at the point of discharge ($mg\,O_2/l$), D_0 the dissolved oxygen deficit at the point of discharge ($mg\,O_2/l$) and D_t the dissolved oxygen deficit at time t (days).

This equation may be used to predict the oxygen deficit of a watercourse at any given point below a discharge, assuming that the ultimate BOD of the waste is known and that values of the first-order reaeration and deaeration constants have been computed.

An equation to predict the value of the critical oxygen deficit is also obtained from equation (1.25). Inspection of Figure 1.13 reveals that at the critical point there is a turning-point where the reaeration rate is equal to the deaeration rate, i.e.

$$\frac{dD}{dT} = 0 \tag{1.28}$$

thus substituting this in equation (1.25) yields:

$$D_c = \frac{k_1 L_0 \exp(-k_1 t_c)}{k_2} \tag{1.29}$$

where D_c is the value of the critical oxygen deficit. In order to solve this equation numerically, it is necessary to have a value for the time at which the critical oxygen deficit is reached. This may be obtained by differentiating equation (1.27) with respect to t and setting dD/dt as equal to zero.

$$t_c = \frac{1}{k_2 - k_1} \ln \frac{k_2}{k_2} \left[1 - \frac{D_0(k_2 - k_1)}{L_0 k_1} \right] \tag{1.30}$$

Limitation in the Streeter–Phelps equation

Although the Streeter–Phelps approach to modelling oxygen profiles in a watercourse provides a simple and extremely useful tool, it is based on a number of assumptions which, under certain circumstances may prove erroneous and lead to large errors. The two assumptions which are likely to produce the largest errors are that wastes discharged to a watercourse are evenly distributed over the river's cross-section and that they then travel down the river as a plug, experiencing no mixing along the axis of the river. In addition, equation (1.24) assumes that only two processes contribute to changes in oxygen concentration, i.e. oxygen is removed along a stretch by microbial oxidation of the added BOD, and it is replaced by reaeration at the surface.

The assumptions regarding the distribution of waste discharges within a water are quite reasonable for the majority of watercourses, once the waste has travelled a reasonable distance downstream and has been distributed across the river. However, it is unlikely to prove valid at the point of effluent discharge since the effluent travels as a plume for some distance downstream before becoming mixed. As regards the addition and removal of oxygen, however, many other processes are occurring which influence this and these include:

1. The removal of BOD by sedimentation. This is particularly important when a significant component of the wastewater BOD is in the form of suspended solids.

2. The conversion of BOD present as suspended solids in the benthal layer into soluble BOD which passes into the water above.

3. Removal of oxygen by the benthal layer to sustain the reactions taking place in (2).

4. Additional inputs of BOD along the stretch resulting from local run-off.

5. Large diurnal variations in oxygen removal and addition resulting from algal photosynthesis and respiration.

Consequently, care is needed in the determination of k_1 and k_2.

Determination of k_1 and k_2 values

The main problems encountered with the application of the Streeter–Phelps equation lies in the determination of accurate values for the reaeration and deaeration constants. The deaeration constant of a particular sample is generally calculated from analysis of data obtained from BOD experiments. A number of techniques are available for this including the method of least squares and the Thomas graphical technique. A simplified technique for use on watercourses is to use the equation:

$$k_1 = \frac{1}{t}\log\frac{L_a}{L_b} \qquad (1.31)$$

where t is the time of travel in days between two stations a and b, and L_a and L_b represent the BOD measured at each of these stations.

It is possible to calculate the reaeration rate constant directly from the Streeter–Phelps equation providing that the other relevant information is known. However, values calculated in this way tend to be low, as errors made in the calculation of the time of water travel, dissolved oxygen concentration, BOD_u and k_1 will all be reflected in determination of k_2. Several empirical formula are available for determination of k_2 based on an analysis of the stream profile; one simple version takes the form:

$$k_2(20\,^{\circ}\mathrm{C}) = 2.833\frac{V}{H^{3/2}} \qquad (1.32)$$

where V is the mean stream velocity (m/s) and H the mean depth (m).

The figure 2.833 is a factor to correct for changes in the surface roughness in the bed of a watercourse which will affect the velocity profile of a stream. Many other equations have been developed for the estimation of this constant which vary in complexity and yield different approximations. The difficulty lies with the large number of factors known to influence the magnitude of k_2. These include:

1. The presence of algae which alter the dissolved oxygen concentration diurnally through their cycle of photosynthesis and respiration.

2. The presence of pollutants such as oil and detergents which alter the physicochemical properties of the air/water interface and thus interfere with oxygen transfer.

3. The presence of organic pollutants on the bed of the watercourse which exerts an oxygen demand that is not detected during determination of the BOD_u of the water.

The effects of the oxygen sag on aquatic life

The saturated dissolved oxygen concentration which a watercourse can support is dependent upon the temperature, but ranges from 12.8 mg O_2/l at 5 °C to 7.7 mg O_2/l at 30 °C. The prevailing oxygen concentration proves to be one of the strongest selection pressures in the determination of the abundance and distribution of the aquatic community, and many vertebrate and invertebrate species have very fastidious oxygen requirements. Indeed many organisms are quite capable of living in waters with a heavy organic load, providing that an adequate dissolved oxygen concentration is maintained (for instance by strong currents, waterfalls or weirs). Thus the oxygen sag profile along the length of a watercourse will be mirrored by the composition of the community within the watercourse. It is observed that when a water is polluted with an effluent there is usually (Figure 1.12):

1. A fall in the total number of species of organisms;
2. A change in the type of species present;
3. A change in the numbers of individuals of each species in the water.

These changes result from either the death of those organisms which are sensitive to the decreased dissolved oxygen concentration, or from their moving away to sites where conditions are more favourable. This allows a recolonisation of the site by organisms capable of tolerating the reduced oxygen levels, but which would previously have been eliminated by predation. This observation has a practical significance as the presence of intolerant or sensitive organisms indicates unpolluted water, whereas their absence coupled with the presence of tolerant organisms indicates polluted water. In addition, changes in community composition may be monitored regularly in order to indicate the long-term environmental effects of wastewater discharges, which would not be apparent from chemical assessments such as BOD and dissolved oxygen concentration. The invertebrate community of a watercourse has proved the most useful indicators for such a scheme because in addition to their sensitivity, they are routinely found in and among the stones, mud and sand on the water bottom, thus making sample collection very easy (Figure 1.13). The resultant correlations between presence or absence of characteristic species and water quality is known as a biotic index. A more sensitive method, known as a diversity index, looks at both the number of species and the number of individuals of each species in a sample.

Obviously the exact species changes which occur with changes in oxygen profile is very geographically dependent. A typical biotic index is illustrated in Table 1.6 which was derived for use on the River

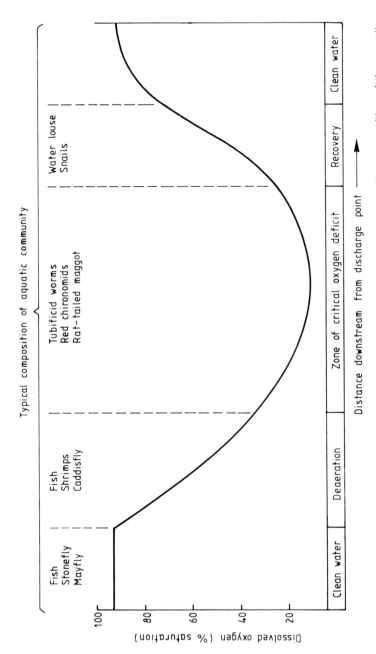

Figure 1.12 The effects of the oxygen sag resulting from a point source discharge, on the composition of the aquatic community.

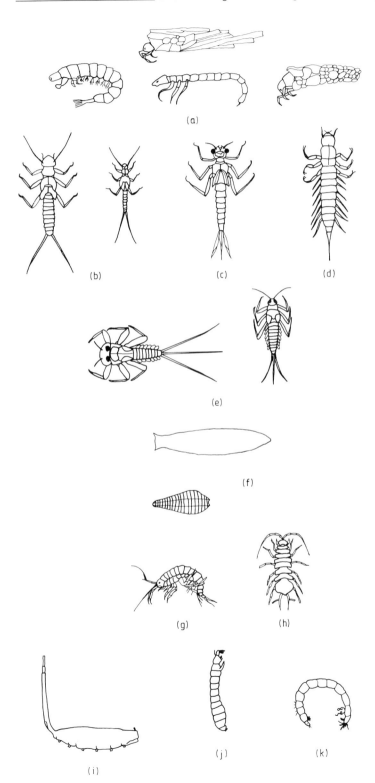

Figure 1.13 Typical invertebrate species which are used to compile biotic and diversity indices: (a) caddis-fly larvae, (b) stonefly nymphs, (c) dragonfly nymph, (d) alderfly larva, (e) mayfly nymphs, (f) freshwater leaches, (g) freshwater shrimp, (h) freshwater hog louse, (i) rat-tailed maggot, (j) *Simulium* pupa, (k) chironomid larva.

Groups (families)	Score
Mayfly larvae (e.g. Euphemeridae, Ecdyonuridae) Stone-fly larvae (e.g. Leuctridae, Perlidae) Cased-caddis larvae (with stones and sand)	10
Damselfly and dragonfly larvae Caseless (free-living) caddis larvae (e.g. philopotamidae)	8
Mayfly larvae (Caenidae) Cased-caddis larvae (with plant debris)	7
Large freshwater mussels (60 × 100 mm) (Unionidae) Freshwater shrimp (Gammaridae)	6
Water bugs (e.g. Corixidae) Water beetles (e.g. Haliplidae, Elminthidae) Caseless caddis larvae (e.g. Hydropsychidae) Fly larvae (e.g. Tipulidae, Simulidae) Flatworms (e.g. Planariidae)	5
Mayfly larvae (Baetidae) Alderfly larvae (Sialidae) Water mites	4
Snails (e.g. Lymnaeidae, Planorbidae, Physidae) Small freshwater bivalves (e.g. Sphaeriidae) Leeches (e.g. Glossiphoniidae) Water hog louse (Assellidae)	3
Fly larvae (midge) (Chironomidae)	2
Worms (e.g. Tubificidae) Fly larvae (rat-tailed maggot)	1

Table 1.7 Simplified version of a typical biotic index.

Trent in the UK. While this particular table applies exclusively to that river community, the principles of the scheme are universally applicable, and similar tables may be derived for any watercourse in any country.

2 Conventional Wastewater Treatment Processes

2.1 INTRODUCTION

Engineers who are required to design wastewater treatment plants are faced with a large choice of individual treatment options (or unit processes). If these are selected wisely and appropriately, then they are capable of producing an effluent which meets any given discharge standard. It is quite feasible to design a treatment plant which discharges an effluent having no ecological impact on a receiving watercourse. However, in any country there are many competing demands for development funds and wastewater treatment projects do not always have a top priority. Thus designers are frequently called upon to meet rigid environmental constraints while restricted by financial limitations. It is important therefore for design engineers to understand the relationship between prevailing environmental conditions (such as climate, land availability and ready availability of chemicals) and the ability of a particular unit process to perform to its design specification. This is of particular relevance in developing countries where it is important to ensure that money is spent to produce the greatest cost benefit. Developing countries pose additional problems in that many facilities which are taken for granted in the developed countries are simply not available. This includes such things as the ready availability of skilled personnel, a guaranteed and dependable source of power and the availability of facilities for plant operation and maintenance.

All wastewater treatment plants are required to reduce the level of suspended solids and organic material in the inflowing sewage; in addition, however, many plants are also expected to remove nutrients and demonstrate a high removal efficiency for pathogenic micro-organisms. No single unit process is currently available which can successfully and efficiently achieve all of these requirements and consequently a combination is required. The remainder of this chapter is concerned with an appraisal of some of the more common unit processes along with recommendations as to their advantages and limitations.

2.2 PRELIMINARY TREATMENT

Screening

Raw sewage delivered by the sewers to a treatment works will contain appreciable amounts of floating materials (such as wood, rags, paper and faecal material), as well as heavier solids such as grit and large suspended solids. It is necessary to remove these at an early stage in the treatment process in order to prevent damage to mechanical equipment such as pumps and aerators, and also prevent the blockage of pipes and valves. The removal of these materials is known as preliminary treatment. It is a simple operation achieved by passing the sewage through a series of screens or strainers. These comprise a set of inclined parallel bars fixed a certain distance apart, which for inlet screening is usually 12, 18 or 24 mm. The inlet screen bars are generally fixed in position and cleaned by a raking mechanism which clears the bars and discharges screenings into an appropriate container (Figure 2.1).

The velocity of flow to the screens should always exceed 0.5 m/s to prevent sedimentation of solids and this is accomplished by choosing an appropriate channel width according to the equation:

$$W = \frac{Q}{VD} \times \begin{cases} 2. & \text{(for 12 mm spaces)} \\ 1.67 & \text{(for 18 mm spaces)} \\ 1.5 & \text{(for 24 mm spaces)} \end{cases} \quad (2.1)$$

where W is the width of channel (m), Q the flow rate (m³/s), V the velocity through the bars (m/s) and D the depth of flow (m).

Passage through the screens will result in a slight head loss, and this is given by

$$HL = \frac{V_1^2 - V_2^2}{2g} + \frac{0.5\,V_2^2}{2g} \quad (2.2)$$

where HL is the head loss (m), V_1 the velocity between bars (m/s), V_2 the velocity in approach channel (m/s) and g the acceleration due to gravity (m/s).

A minimum of two sets of screens are generally provided in parallel at all but the smallest works to allow one set to be taken out for maintenance. In addition, a bypass channel often proves a useful addition in the event of a screen blockage occurring.

The discharged screenings form an extremely unpleasant and potentially hazardous material. They are generally disposed of in landfill sites or by incineration. An alternative method is to pass them through a macerator and return them to the inlet flow of the works such that they are removed by primary sedimentation. This process may be combined with the screening process by means of a comminutor.

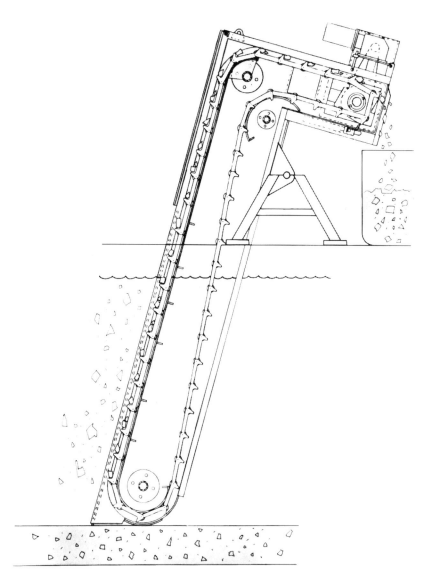

Figure 2.1 Mechanically raked inlet screen bars. (Courtesy of Jones and Atwood Ltd, Stourbridge, UK.)

Comminutors

These are sited in place of bar screens and comprise a rotating drum with attached cutting teeth. The rotation of the drum causes the cutting teeth to pass through fixed combs which retain the larger pieces of sewage. These are thus macerated and pass through the drum into the downstream flow (Figure 2.2). Although comminution removes the need for screening disposal, it does place an additional burden on the primary sedimentation tanks. It is also associated with

Figure 2.2 Typical example of a comminutor. (Photograph courtesy of Jones and Atwood Ltd., Stowbridge, UK.)

a larger loss of head than screens, which may be an important point at works where the available head is limited. In addition the material which has passed through a comminutor has a tendency to re-form and agglomerate into large balls during passage from the comminutor to the primary sedimentation tanks. These are frequently of sufficient diameter to cause pipe blockages.

The important role that efficient screening has at a sewage

(a)

(b)

Figure 2.3 (a) Blocked nozzles on a trickling filter distributor arm; (b) an extreme solution to a blocked filter arm, in this example the end of the arm has been broken off and the sewage flows down the draught tube without receiving any biological treatment.

treatment works is often overlooked, particularly when designing for developing countries. One of the most frequent reasons for the failure of trickling filters in developing countries is due to blockage of the nozzles in the rotating distribution arms by materials which should have been removed by screening (Figure 2.3). Although regular maintenance would circumvent this problem, it is very rarely available. Consequently, screens should be selected which require a minimum of skilled maintenance, and then only on a routine basis. In addition they should possess a minimum of working parts for which spares are available off the shelf from within the country of operation.

Grit removal

Efficient grit removal is extremely important at plants in order to prevent abrasion and wear of mechanical equipment, deposition of grit in pipes or channels and accumulation in aerators and anaerobic digestors. A well-designed grit chamber should remove 95% of particles with a diameter > 0.2 mm. Poor grit-chamber operation results in the removal of grit which has a high organic content. Grit removal is achieved by the creation of a zone in which the velocity is constant at 0.3 m/s, under which conditions grit will be deposited but other suspended organic material will not. This is nearly always achieved by means of a chamber which is generally parabolic, trapezoidal or V-shaped in cross-section. This is matched to a rectangular, critical-depth control flume and the flow is controlled by a standing wave flume at the outlet end. The channel width for these chambers is given by

$$W = \frac{1.5 A}{H} \tag{2.3}$$

where W is the width (m), A the area of flow (m^2) and H the depth of flow (m).

The majority of grit particles will settle at a velocity of 0.03 m/s, and since the velocity in the channel is 0.3 m/s then a channel of ten times the depth of sewage should ensure that the majority of grit has settled out. In practice it proves necessary to design the channel length twenty times the depth of sewage in order to compensate for turbulence. The deposited grit is removed from the channels either by suction pumps or by bucket dredgers mounted on moving bridges.

Aerated grit chambers

An alternative to the constant velocity chamber is provided by aerated grit chambers in which air is used to create a spiral flow pattern through the chamber. The intensity of motion permits grit to settle while retaining the organics in suspension (Figure 2.4). It is thus not the flow through velocity, but the roll velocity of the air which controls the quantity and quality of grit removed, and this is controlled by the rate of air supply. Typical roll velocities in such chambers are 0.25–0.6 m/s at the surface and 0.03–0.3 m/s at the bottom. The shape of aerated grit chambers is not of primary importance but rather the provision of uniform flow patterns and hydraulic control to minimise short-circuiting. This may be best achieved with a steady air source, adequate baffling and proper placement of the air diffuser heads. A retention time of 3–5 minutes is generally found adequate under these conditions. The settled grit is conveyed to a hopper generally by means of a tubular conveyor.

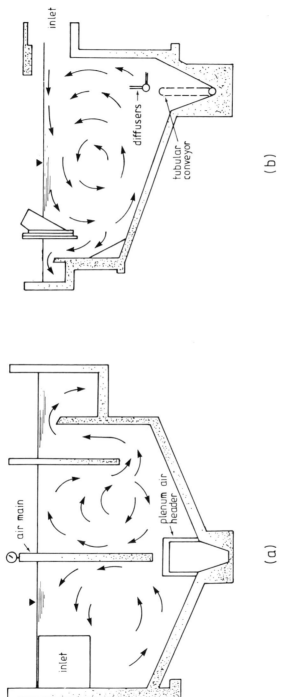

Figure 2.4 Cross-section through typical aerated grit chambers: (a) air header placed centrally and grit removal by pump; (b) air header placed laterally and perpendicular to the flow, grit removal by chain and bucket.

2.3 PRIMARY SEDIMENTATION

Introduction

Sewage which has been subjected to preliminary treatment has had the majority of large solid and floating material removed. However, it will still contain a high concentration of suspended particles in the size range 0.05–10 mm, and these are know as settleable solids. It is not essential for settleable solids to be removed before biological treatment, and indeed many biological unit operations are designed to operate without the provision of primary sedimentation. However, a well-designed sedimentation tank can remove 40% of the BOD in the form of settleable solids. The advantages of this are that it reduces the BOD load to the next stage of treatment and thus permits smaller reactor sizes, with consequent lower power consumption. In addition, the reduced BOD loading results in a smaller surplus sludge production, which in turn allows the provision of smaller secondary sedimentation tanks. As a result of these advantages, sedimentation is the most widely used process in wastewater treatment and, with the exception of waste-stabilisation ponds, are the most cost-effective units at a wastewater treatment plant.

All sewage treatment works should be provided with at least two primary sedimentation tanks such that one may be removed from service for repair and maintenance without affecting the sediment-ation process. The actual number will of course depend on the flow rate, but where more than one are operated consecutively, it is normal practice to operate these in parallel. This provides problems, however, with flow distribution as it is difficult to ensure that each tank receives both an equal flow rate and an equal solids loading. To reduce these discrepancies as much as possible, some degree of mixing should be provided before flow division takes place. In addition, it is important to ensure that the sewage velocity is self-cleansing during distribution to prevent solids accumulating within the distribution system.

Types of sedimentation tanks

An ideal sedimentation tank should aim to provide four important facilities:

1. An inlet zone which facilitates dissipation of the flow velocity of the incoming sewage. This should be achieved as rapidly as possible to ensure that the inlet zone occupies a minimum of space. If the inlet is badly designed, then excessive flow velocities may be experien-ced in the tank, resulting in a reduction in settlement efficiency due to shortcircuiting.

2. An outlet weir to collect the settled sewage prior to further

distribution. This is generally a V-notched weir around the tank periphery that allows liquid to be withdrawn as a thin layer. These weirs frequently have a scum board installed in front of them 75–150 mm above and 300–400 mm below the surface of the sewage. These prevent floating material, oil and grease from passing over the outlet weirs. It is necessary therefore to provide facilities to allow this floating material to be removed frequently to ensure that it does not accumulate and pass over the scum board.

3. A settlement zone which represents the actual tank capacity. It is in this area that settlement takes place, and it should be free from short-circuiting and stagnant areas.

4. A zone for the storage, collection and withdrawal of sedimented solids. This may either be manual or more usually mechanical. Tanks which employ manual desludging will obviously require sufficient capacity to store the sludge between desludging. This, together with the unpleasant nature of the desludging process, means that such tanks are rarely if ever installed at new treatment works.

Three types of sedimentation tanks are routinely employed at treatment works and these are designated according to their flow regime.

1. Horizontal flow tanks.   Such tanks are usually rectangular in plan with a length/breadth ratio of up to 4 to 1. Sewage flows continuously from an inlet at one end to an outlet at the other. The heavier solids are therefore desposited at the inlet end, with the lighter solids being deposited towards the outlet end. These tanks may be desludged manually, or more usually mechanically in which case the floor has an inverted pyramid-shaped hopper at the inlet end. The sludge is pushed into the hopper by means of a scraper driven along the floor of the tank (Figure 2.5).

2. Radial flow tanks.   These tanks are circular in plan with a floor which slopes down to the tank centre. The sewage enters at the centre of the tank and flows up and out to the tank perimeter which is fitted with an outlet. The velocity of the sewage is thus greatest at the tank centre, where only the heavier particles settle out, and decreasing towards the perimeter, allowing sedimentation of the smaller particles. Sewage flow in such tanks is thus subjected to both upwards and radial flow patterns. The solids which settle on the floor are removed mechanically by means of revolving scrapers which channel the sludge into a central hopper. Although such tanks cannot be built with common walls like rectangular tanks, they are much shallower than upward-flow tanks and thus offer large savings in excavation costs.

Figure 2.5 Mechanically desludged, horizontal-flow primary sedimentation tank. (Photograph courtesy of Biwater Treatment Ltd., Heywood, UK.)

3. Upward-flow tanks. These may be either circular or square in plan with a sharply sloping floor and can be viewed as an inverted cone. The sewage inlet is at the tank centre, below the water line but above the sludge blank, and the direction of flow is initially downwards. However, this then changes direction to become outwards and upwards towards the outlet weir, set around the tank periphery. As the flow direction is upwards and the solid particles are attempting to settle downwards, a point is reached at which the particles are stationary, permitting aggregation and the formation of a sludge blanket. Such a mechanism permits an increased solids removal efficiency and high effluent quality. Because of the steep sides of these tanks (which should make an angle of not less than 60° to the horizontal), sludge falls to the bottom of the hopper and may be withdrawn via a pipe set at the base of the tank (Figure 2.6).

It has been observed in practice that the shape of the tank has little effect on the upward-flow velocity in the tank, or in the concentration of sludge removed from the tank. However, circular tanks appear to

Figure 2.6 Upward-flow primary sedimentation tank. (Photograph courtesy of Biwater Treatment Ltd, Heywood, UK.)

be more efficient at the removal of solids and BOD, and should thus be chosen preferentially when site layout permits.

Inlet configuration

The single most important factor influencing the settlement of solids is local velocities, and these should be kept below 15 mm/s in order to achieve >50% solids removal. Increasing flow rates are one of the main reasons for increased local velocities as a result of increased turbulence. Inlet configuration is thus a key element in tank design as it is responsible for a uniform distribution of liquid and solid flow across the tank. Wastewater is usually delivered to the tank by pipes, manifolds or open channel troughs. Under ideal conditions each outlet should be spaced evenly and discharge equal amounts of solid and liquid. In the case of rectangular tanks, where wastewater flow approaches the tank at right angles, an unequal flow is generated through the outlet ports. Consequently symmetrical flows are preferred whereby the flow enters parallel to the tank and symmetrical to its centreline. This arrangement leads to a more uniform flow distribution. Where asymmetric inlet configurations have been installed they may frequently prove to be the cause of poor tank operation. In these situations a more even flow distribution is often achieved by creating a higher head loss in the inlet ports relative to that in the trough.

For both inlet arrangements it is essential to provide a minimum inlet trough velocity of 0.6 m/s at average flow and 0.3 m/s at minimum flow in order to prevent solids deposition. If such velocities cannot be achieved (for instance owing to unanticipated changes in wastewater flow), then alternatives such as mixing by channel aeration, or installation of water-jet nozzles in the inlet trough may be contemplated. These measures will of course have the drawback of high maintenance and operating costs.

2.4 SECONDARY TREATMENT

Introduction

The aim of secondary treatment is to reduce the oxygen demand of an influent wastewater to a given level of purification. This is generally such that on discharge to a receiving water the treated waste will have no detrimental effect on the environment. Under certain circumstances, however, particularly for industrial wastes, the aim may simply be to reduce the pollution potential of the wastewater without incurring excessive costs. A large number of biological unit operations are available to achieve the aerobic oxidation of BOD. These all have two essential features in common, in that a reactor is required to contain the organisms which will effect the BOD removal and a mechanism is required to ensure that there is always a residual oxygen concentration. All unit operations can be classified on the basis of their microbial population, into either fixed film or dispersed growth processes. In the former, as the name suggests, the microorganisms are immobilised or attached to an inert support which is maintained in contact with the inflowing sewage. In the latter the microorganisms and wastewater are kept in intimate contact by mixing, the mixing apparatus also being responsible for keeping the suspension aerated.

(A) Fixed film processes

Fixed film processes represent the oldest form of wastewater treatment system. Land application of sewage is a type of fixed film system whereby the microorganisms are attached to soil particles and the wastewater percolates down through the soil, undergoing purification during its passage. Land application of sewage was widely practised in the UK and led to the development of the trickling or percolating filter in the nineteenth century. Trickling filters have been used extensively world-wide since that time and represent a simple and reliable form of effluent treatment. During the last 30 years, a modification of the process known as a rotating biological contactor (RBC) has started to prove popular, particularly in Japan and the USA.

1. Trickling filters

A trickling filter (also referred to as a biofilter, bacteria bed or percolating filter) is a reactor of rectangular or circular plan which is filled with a permeable media. Wastewater is distributed mechanically over the media and percolates down the filter to collect in an underdrain system at its base. A microbial film develops over the surface of the media and this is responsible for removal of BOD during passage of sewage through the bed. Filtration, in the traditional sense of straining out of large particles, does not occur and the process is solely a biological one. The efficiency of the system depends upon an even distribution of settled sewage over the whole surface of the filter, and also upon the circulation of air throughout the filter media. On circular filters the former is achieved by influent entering at the centre of the bed and passing into radial distributor arms above the bed surface. The distribution arms are fitted at intervals with sparge holes such that discharge of sewage through them provides the necessary force to drive the arms around the central column (Figure 2.7). In the case of rectangular filters the

Figure 2.7 Distribution of sewage around a trickling filter by means of a rotary distributor arm.

distributor is driven forward and backwards with the liquid being siphoned from a channel running along the length of the bed (Figure 2.8). This wetting followed by a rest period is an essential requirement for successful filter operation. Blockage of the sparge holes is a regular occurrence and regular cleaning is essential to maintain optimum filter performance. Badly maintained filter arms are the commonest cause of filter failure in developing countries and many ingenious methods have been adopted by local operators in an attempt to avoid this routine chore (see Figure 2.3).

The distribution of air through a filter occurs by circulation through void spaces in the filter media. This is encouraged by temperature differences between the air and wastewater which causes an up-draught of air through draught tubes located in the sides of the filter. The filter media must therefore provide both a large surface area for the growth of a slime layer and also a large void space for oxygen transfer and liquid flow. Traditionally, crushed rock or blast-furnace slag of diameter 25–100 mm has been employed, but this has limited filter depth to 3.0 m. More recently the development of lightweight plastic media has permitted filter depths of up to 12 m

Figure 2.8 Distribution of sewage across a rectangular trickling filter. (Photograph courtesy of Biwater Treatment Ltd, Heywood, UK.)

to be exploited. These have proved particularly popular in the treatment of high-strength industrial wastes.

The recommended loading rates for filters will depend upon the effluent quality required, the media employed and the type of wastewater being treated; typical figures are given in Table 2.1. Loadings in excess of quoted figures cause excessive growth of the slime layer resulting in the voids between the media becoming blocked. This prevents sewage percolating through the filter and thus it collects on the surface of the media. When this happens the filter is said to have 'ponded'. In a well-operated filter there is a continual cycle of film growth followed by death and detachment from the media. This is known as 'sloughing off' and the resultant sludge is carried away with the filter effluent. For this reason a sedimentation tank is required to settle out and remove solids from filter effluents and these are known as humus tanks, the resulting sludge being referred to as humus sludge. Periodic sloughing off the slime layer is a feature of filter operation and may be expected at the change of seasons when warmer weather returns. During this period treatment will be severely reduced, although it is rapidly re-established as a new slime layer grows.

The major advantage of trickling filters is that they are comparatively simple to operate and have very low running costs. In addition they are able to tolerate shock and toxic loads owing to the short contact time of the wastewater with the slime layer. They do have many disadvantages, however. Their land requirements are high and they can only provide a limited treatment efficiency (although a 20/30 effluent is readily attainable). In hot countries they are associated with odours and fly nuisance, and although it is a relatively simple technique the degree of skill required for operation and maintenance is frequently not available.

Parameter	Low rate	Intermediate rate	High rate
Depth (m)	1.5–3.0	1.25–2.5	1.0–2.0
Recirculation ratio	0	0	1–3; 2–1
Organic loading $(kg/m^3 d)$	0.08–0.32	0.24–0.48	0.32–1.0
Hydraulic loading $(m^3/m^2 d)$	1–4	4–10	10–40
Effluent quality	Fully nitrified	Some nitrification	Little nitrification

Table 2.1 Process characteristics of trickling filters with mineral media.

2. Rotating biological contactors (RBC)

Although the idea for an RBC was patented as early as 1900, it was not until the 1960s that this process was employed commercially. An RBC (also referred to as a biodisc) operates on a similar principle to the trickling filter, but has a rotating bed of attached bacteria which is immersed in a tank of wastewater. The rotation exposes the surface of the disc to the atmosphere permitting aeration, and then resubmerges it in the wastewater. The bed comprises a number of circular discs, closely spaced together and mounted on a rotating drive shaft (Figure 2.9). The discs rapidly develop a microbial community in the form of a film up to 3 mm thick and this film is responsible for BOD removal. As the film accumulates, sloughing off occurs in an identical manner to trickling filters. The discs may be made of wood, metal or polystyrene in either a flat, corrugated or honeycombed profile in order to increase the surface area for film growth. Full-scale RBCs may have discs up to 4 m in diameter mounted on 7 m shafts. Rotation speeds are usually between 1 and 2 rpm. Rotation of the discs plays several roles in that, as well as aeration, it also keeps the tank contents mixed and helps produce the shear forces necessary for sloughing off.

The major advantage of an RBC is its ease of operation and low land requirement. The former suggests that it may be suitable for many applications in developing countries and within northern Europe and the USA it has found application in rural areas at

Figure 2.9 RBC drive shaft, illustrating the circular discs which accumulate the microbial film. (Photograph courtesy of Biwater Treatment Ltd, Heywood, UK.)

unattended sites. It is a particularly suitable process for marketing as a package plant for small communities. In addition to these, several other advantages have been quoted which include:

1. Reduced power and maintenance costs;

2. Stability against hydraulic shock loadings;

3. Capability of achieving a high degree of carbonaceous and nitrogenous BOD removal;

4. Ponding of the bed and clogging of filter nozzles is eliminated.

Of course RBCs are not without their disadvantages, the majority of which seem to stem from the fact that it is still a relatively novel process. Thus several problems have been encountered such as breaking or cracking of the discs and regular shaft bearing failure. The RBC process is generally equipped with an enclosure to protect it from rain and wind which, if badly designed, may hinder access for inspection, maintenance and repair of the discs. Perhaps the most quoted disadvantage of the RBC process is its lack of operational control, the only control options available being the drive-shaft speed and the depth of immersion of the discs. However, in the context of its applicability in developed countries, such simplicity of operation should be regarded as an asset, providing of course that the appropriate unit has been supplied in the first place. Consequently the onus for successful operation lies in the hands of the designer as there is no flexibility within the system to compensate for design error.

(B) Dispersed growth processes

The activated sludge process

The activated sludge process is a suspended growth system comprising a mass of microorganisms constantly supplied with organic matter and oxygen. The microorganisms grow in flocs, and these flocs are responsible for the transformation of the organic material into new bacteria, carbon dioxide and water. The flocs are constantly being washed out of the reactor to the secondary sedimentation tank by the flow of incoming sewage. Here they flocculate and settle under quiescent conditions. It is a characteristic of the activated sludge process that a fraction of this settled sludge is recycled back to the aeration tank in order to provide sufficient biomass to achieve efficient BOD removal. The remainder of the solids are wasted. The fraction of the solids which are wasted determines the average amount of time which a microorganism will spend in the reactor. This is termed the sludge age and is defined as

$$\text{Sludge age (d)} = \frac{\text{Total solids in reactor (kg)}}{\text{Total solids wasted (kg/d)}} \qquad (2.4)$$

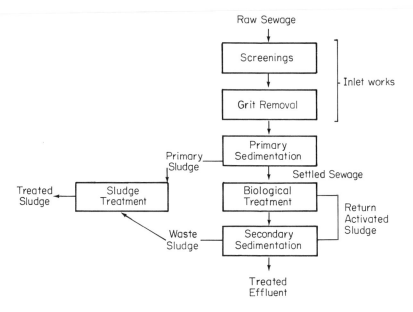

Figure 2.10 Schematic diagram of a conventional activated sludge plant.

A typical schematic diagram of a conventional activated sludge plant is shown in Figure 2.10.

There are a large number of variations on the conventional activated sludge process, however, these vary only in three principal ways:

1. The method of oxygen supply;

2. Loading rate;

3. Aeration tank configuration.

Methods of oxygen supply.   The two techniques available to aerate activated sludge reactors employ either mechanical aeration or diffused air.

(a) Mechanical surface aeration.   These devices transfer oxygen from the atmosphere to the mixed liquor by causing agitation at the liquid surface. This produces numerous fine droplets which are in contact with the atmosphere. Thus the liquid surface area available for gas transfer is increased and a high concentration of air is entrained. As these saturated droplets fall back on to the liquid surface they are dispersed around the tank by the aerator. In addition this agitation helps to keep the influent and sludge in suspension and prevents the flocs from settling.

Figure 2.11 Vertically mounted mechanical aerator at the centre of one pocket of an activated sludge reactor. (Photograph courtesy of Biwater Treatment Ltd., Heywood, UK.)

Figure 2.12 Horizontally mounted mechanical aerator across the oval channel of an oxidation ditch. (Photograph courtesy of Biwater Treatment Ltd., Heywood, UK.)

Aerator type	Standard oxygen transfer rate (kg O_2/kWh)
Diffused air	
Fine bubble	1.2–2.0
Coarse bubble	0.6–1.2
Jet	1.2–2.4
Pure oxygen	
UNOX	2.4–3.8
VITOX	2.8–4.2
Mechanical	
Simcar surface aerator	2.1–2.4
Turbine aerator	2.1–3.2
Simplex cone	2.0–2.6
Oxidation brushes	
Kessener brush	2.4–3.2
Cage rotor	1.4–3.0

Table 2.2 Standard oxygen transfer rates of typical aeration devices.

Mechanical aerators may be mounted either vertically or horizontally. The former are most common at plants where the reactor is divided into a series of pockets, with an aerator mounted at the centre of each pocket (Figure 2.11). The latter is used almost exclusively in oval channel tanks which form a continuous loop and are known as oxidation ditches (Figure 2.12). The characteristics of the main mechanical aerator types are listed in Table 2.2. The oxygenation capacity of surface aerators may be varied by altering the rotor height above the liquid surface or by varying the level of liquid in the reactor. If rotor depth of immersion is increased beyond the optimum for aeration, then although good mixing is achieved, aeration efficiency is low. The immersion depth for an aerator is very important and its rated capacity will be seriously reduced if it is incorrect.

(b) Diffused air aeration. Diffused aerators are mounted on the floor of a reactor and release air into the mixed liquor as a stream of bubbles. Oxygen transfer takes place at the interface of the air bubble and the water as it rises to the surface. The circulation currents initiated as the bubbles rise to the tank surface are also responsible for keeping the mixed liquor solids in suspension. Two types of air-distribution system are available. Porous tiles made of ceramic or more recently reticulated foam, produce bubbles of diameter up to 2 mm and are known as fine-bubble diffused air. Bubbles of this size provide a large air/liquid interface and thus oxygen transfer efficiency is high. The diffusers do, however, suffer the problem of blockage, either from attached biomass growing around the tiles or from dust particles present in the air. This latter problem is particularly

prevalent in industrial areas and hot dusty climates where the air has a high dust content. To alleviate this problem, coarse bubble aerators are available in which air is pumped through holes in pipes laid along the floor of the reactor. This produces bubbles with diameters in the range 3–5 mm with transfer efficiencies reduced accordingly.

Loading rate.   Four basic parameters are available to describe the rate at which settled sewage is applied to an activated sludge reactor, all of which are loosely termed 'loading rate'.

   1. Volumetric loading.   This parameter determines the amount of time which the sewage will undergo aeration in the reactor and is defined as

$$\text{Retention time (d)} = \frac{\text{Reactor volume (m}^3)}{\text{Total daily flow (m}^3/\text{d)}} \qquad (2.5)$$

This value is also known as the hydraulic retention time, and in the absence of sludge recycle it has the same value as the sludge age. The aeration tank capacity must be of sufficient magnitude to allow adequate contact time between the wastewater and sludge flocs. This generally requires at least 5 hours to guarantee a 20/30 effluent at dry weather flows. Conventional activated sludge plants have volumetric loadings in the range 6–10 hours.

   2. Organic loading.   The organic or BOD loading is a measure of the amount of BOD which is applied per unit volume of aeration tank capacity and is defined as

Organic load (kg/m^3 d)

$$= \frac{\text{Influent flow (m}^3/\text{d)} \times \text{BOD of sewage (kg/m}^3)}{\text{Reactor volume (m}^3)} \qquad (2.6)$$

   The organic loading a plant receives will vary considerably during the day owing to diurnal fluctuations in both influent flow rate and strength. Determination of daily organic loads cannot therefore be determined accurately from the BOD value of a single influent sample. More accurate values will be obtained from a composited sample comprising 24 1-hourly samples of equal volume. Occasionally where the BOD of the sewage shows particularly wide variations in strength, the volume of the sample should be flow weighted to ensure accurate loading rates.

   3. Food:Microorganism ration (F/M).   The F/M ratio or sludge loading rate is the most useful of the loading parameters. It is defined as

$$\text{F/M ratio} = \frac{\text{BOD of sewage (kg/m}^3) \times \text{Influent flow (m}^3/\text{d)}}{\text{Reactor volume (m}^3) \times \text{Reactor solids (kg/m}^3)} \qquad (2.7)$$

It is the only form of loading over which the operator has any control since increasing the sludge wastage rate will cause an increase in the F/M ratio as the mixed liquor suspended solids (MLSS) will decrease. Decreasing the wastage rate will have the reverse effect. The magnitude of the loading rate will influence the biological behaviour of the sludge in many ways including BOD removal rate, sludge settleability and degree of nitrification.

4. Floc loading. A more recent term has been coined to give an indication of the BOD concentration which is available to the sludge microorganisms at a given time. Unlike the F/M ratio it contains no units of time, but is an indication of the instantaneous substrate concentration at the point of mixing of sludge and MLSS. It is defined as

Floc loading (kg BOD/kg MLSS)

$$= \frac{\text{Mass of BOD at time of mixing}}{\text{Mass of MLSS at time of mixing}} \qquad (2.8)$$

Floc loading is rarely used, but finds application at plants where sludge settlement in the secondary sedimentation tanks is poor. It has been observed that provision for a zone of high floc loading (> 100 kg BOD/kg MLSS) within a plant frequently results in a dramatic improvement in sludge-settling properties.

Effect of loading rate on plant performance. The rate at which a plant receives influent BOD will have a marked effect on effluent quality, sludge-settling properties and the efficiency of plant perfor-mance. Changes in loading rate also affect directly the solids retention time (or sludge age) of the process as they are related by the equation:

$$\frac{1}{\Theta} = Yq - k_d \qquad (2.9)$$

where Θ is the sludge age (d), Y the sludge yield (kg sludge produced/kg BOD utilised), q the organic loading rate (kg BOD/kg MLSS d) and k_d the decay coefficient (d).

This equation is discussed in more detail in Chapter 6, but it is apparent that an increase in the organic loading rate will lead to a decrease in sludge age. Different activated sludge operating regimes may thus be identified by their loading rate and sludge age. Three main processes are recognised, namely high rate, conventional and extended aeration. The operating conditions for each of these regimes and the characteristics of their performance are given in Table 2.3.

Process configurations. Activated sludge reactors require provi-sions for the introduction of settled sewage and its removal, together with the mixing of returned sludge. The exact configuration of tank

Table 2.3 Design and performance characteristics of the main types of activated sludge systems[a].

Process	Sludge age (d)	F/M (kg BOD/kg MLSS d)	MLSS (mg/l)	BOD removal (%)	SSVI (ml/g)	HRT (d)
High rate	1–3	0.6–1.8	4000–5000	60–80	120–140	1–3
Conventional	3–8	0.6–0.6	1200–3500	85–90	40–100	6–8
Nitrifying	8–12	0.05–0.2	3500–5000	90–95	40–80	10–16
VITOX	5–25	0.1–0.8	3000–12000	90–95	40–70	1–5
Extended aeration						
Oxidation ditch	60–90	0.02–0.10	3500–5000	90–95	100–200	20–30
Package plant	30–40	0.12–0.50	3500–5000	75–90	100–300	20–25

[a]All data are typically at 15 °C. Higher temperatures give much greater efficiencies.

selected will have a profound effect on many aspects of plant performance and economy. The major types of plant configuration are detailed below.

1. Batch reactors. The original activated sludge process was operated as a batch reactor and known as the fill and draw process. A reactor was filled with settled sewage and aerated for a sufficient time to oxidise the majority of the BOD (generally 8 hours). The reactor contents were then allowed to settle and the treated supernatant discharged to a watercourse. A portion of the settled sludge was wasted and the whole process repeated again. Fill and draw reactors lost favour because of the amount of operator control they required. With the advent of microprocessor control, a recent modification of this process, known as a sequencing batch reactor (SBR), is gaining increased popularity. An SBR allows several processes, such as carbonaceous oxidation, nitrification, denitrification and phosphate removal, to be performed in the same reactor. This technology is particularly appropriate where there are highly variable hydraulic and organic loads and limited skilled operating and maintenance staff. Typical applications might be small communities, vacation resorts and institutional facilities. A typical flow sheet for an SBR is illustrated in Figure 2.13.

2. Complete mix reactors. In complete mix reactors the settled

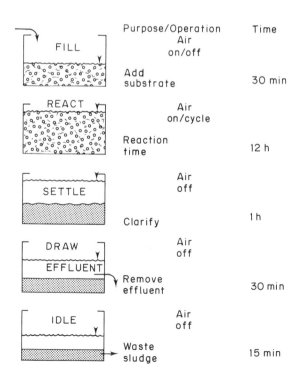

Figure 2.13 Cycle of alternating conditions imposed on the reactor of an SBR.

sewage and return activated sludge are rapidly distributed through-out the tank such that a sample, taken from any point in the reactor, should yield identical values for MLSS, BOD and oxygen con-centration. The advantage of this system is that the large dilution of influent sewage afforded by the aeration tank provides a buffer against any toxic substances which might be present in the influent. In addition there is a uniform distribution of load throughout the tank, which should ensure an efficient use of aerators. Badly designed complete mix reactors may suffer from short-circuiting, however, whereby a fraction of the sewage passes straight through the tank and into the effluent without receiving adequate treatment. In addition, complete mix reactors have a low floc loading and thus sludge-settlement problems are common.

3. Plug-flow reactors. Plug-flow reactors are generally tanks with a high length to breadth ratio. The settled sewage and return activated sludge are introduced at one end and removed at the other. Ideally the sewage particles should pass down the length of the tank without mixing and should emerge in the same sequence in which they entered. If samples are removed longitudinally along the tank, there will be a gradient of decreasing BOD, increasing MLSS and decreasing oxygen demand. In practice a plug-flow reactor consists of a number of tanks or pockets in series, each equipped with its own aerator, thus each pocket behaves as a complete mix reactor. As the load and oxygen demand are not evenly distributed along the reactor, attempts have been made to alleviate this by tapered or step aeration. This modification has a decreasing intensity of aeration along the length of the tank such that the air supply matches the BOD demand and is utilised more efficiently. As the costs of aeration can account for up to 50% of total plant running costs and 50% of total energy costs, then the potential for savings in this area are large. Because there is an area of low MLSS concentration and high BOD concentration at the inlet to a plug-flow reactor, then this represents an area of high floc loading. Problems of sludge settlement are traditionally much reduced with plug-flow reactors.

4. Step feeding. Step or incremental feeding is an alternative method to step aeration with the aim of distributing the load more evenly and thus avoiding oxygen deficiency at the inlet. The technique is employed only on plug-flow reactors and all return sludge is recycled to the reactor inlet. The flow of the incoming settled sewage is then split and fed to a number of pockets. If too many increments of feed are employed, then the reactor will resemble a complete mix system more than plug flow as the gradient of substrate has been reduced. Consequently it is unwise to provide more than two increments or else the disadvantage of complete mix systems are encountered.

Modifications to the conventional activated sludge process. The unifying feature of the activated sludge process is that a portion of the sludge is settled out and recycled back to the aeration tank. However, a large number of variations are possible, based on differing reactor configurations and modes of aeration, but which still provide sludge recycle. These modifications have generally been devised with a specific function in mind, but all of them claim to reduce construction and operating costs if they are correctly applied. Some of them permit modifications to existing conventional plants, whereas others require provision of completely new plant. The choice of the most appropriate type of unit depends, therefore, on an understanding of the philosophy behind the modifications which it incorporates.

1. Contact stabilisation. This process is ideally suited to complete mix plants where problems are encountered with sludge settlement. Contact stabilisation provides a zone with a very high floc loading by the provision of a small tank in which the recycled sludge and screened sewage are mixed. The absorptive properties of the sludge floc are then exploited for rapid BOD removal. The process comprises a small tank in which screened sewage is mixed with return activated sludge for 30 minutes at high loadings and aeration rates. The mixed liquor from this 'contact' zone is then settled, the supernatant discharged to a watercourse and the sludge transferred to a reaeration or stabilisation zone. Here the adsorbed BOD is oxidised by aeration for up to

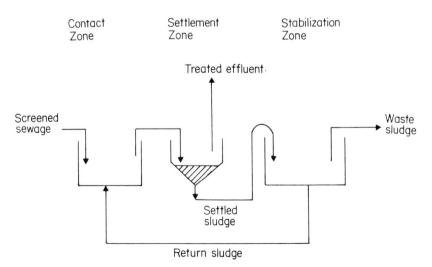

Figure 2.14 Schematic diagram of a contact stabilisation plant suitable for use in a small community.

12 hours and the sludge returned to the contact zone. The main advantage of the process is that there is a large reduction in aeration tank capacity as it is the settled sludge which receives prolonged aeration and not the influent sewage. This process is particularly well suited for supply as a package plant for small communities and industrial wastes. One such plant is illustrated in Figure 2.14. Before specifying the contact stabilisation process for industrial wastes, however, it is important to ensure by laboratory studies that there is a rapid adsorption of BOD by the flocs (85% uptake within 15 minutes is a typical figure). Otherwise the contact step will not prove effective.

2. Oxidation ditches.   Oxidation ditches were originally designed as an inexpensive treatment system to meet the needs of small rural communities. The important features of the original design were that they were shallow (1 m) continuous oval ditches, excavated directly, and with the exception of a simple lining for certain soil types no further civil engineering was required. They operated at very low loading rates, without the need for primary sedimentation. Mixing and aeration was provided by a horizontal paddle aerator spanning the ditch at one point. This also served to keep the mixed liquor in circulation around the ditch (Figure 2.15). After a suitable period of aeration, the aerator was turned off and the ditch served as a secondary sedimentation tank. The low loading rates and long sludge

Figure 2.15 Horizontal paddle aerator operating in an oxidation ditch.

ages meant that sludge production was kept low as a result of endogenous respiration, but periodic removal of excess MLSS from the ditch was required in order to maintain oxygen transfer. The increased land requirements which resulted from the low loadings and long sludge ages was more than offset by the low construction and operating costs. In addition the resulting sludge showed excellent settling and dewatering characteristics, often being fully mineralised. The effluent also is generally better than conventional activated sludge processes. As a result of the long sludge ages, nitrifying bacteria will form a component of the sludge fauna and the effluent is often completely nitrified, incorporation of anoxic zones will also lead to nitrogen removal by denitrification.

Since its inception in the late 1950s, however, oxidation ditches have undergone many modifications and no longer represent a simple wastewater treatment technology. Separate secondary settlement tanks are now almost universal, although these may be sited within the ditch perimeter and thus land requirements are not affected. Alternatively, a side of the ditch may be divided into two channels whereby one channel may be isolated from the main ditch when required and employed as a settlement tank. In addition the ditch is invariably constructed using concrete. Many modifications have been made to the original paddle aerator, and as the size of the ditch is limited by the ability of the aerator to suspend the liquor and propel around the ditch, this has led to increases in ditch dimensions.

Figure 2.16 Vertical shaft cone aerators operating in the Carrousel process.

Horizontal rotors of large diameter and improved blade configuration, known as mammoth aerators, have permitted ditches with depths of up to 3.5 m to be constructed. Vertical shaft cone aerators have also been employed in a modification known as the Carrousel process. Here the aerators are located at the end of baffled tanks which have several channels side by side and connected in series, one aerator for each channel (Figure 2.16). Such a system is primarily for large populations in excess of 10 000 and plants treating sewage with a population equivalent up to 500 000 are in operation.

3. The absorption–bio-oxidation (A/B) process. This system comprises a two-stage activated sludge process operated in series. Its aim is to maximise the production of sludge in the two reactors. This results in a saving in aeration costs, since oxygen is required only for oxidative processes which result in the production of energy and carbon dioxide. The increased sludge production can then be coupled to anaerobic digestion so that the organic material in the sludge can be used for methane production. The process operates without primary sedimentation and is characterised by a separation of the first-stage sludge from that of the second stage, consequently two secondary clarifiers are required. The first or 'A' stage has a very high sludge-loading rate in the range 3–7 kg BOD/kg MLSS d, and this high loading means that the process frequently operates in a facultatively anaerobic mode. Although the treatment efficiency in the A stage is low, a rapid removal of 50% of the influent BOD is achieved and the energy required to remove 1 kg of BOD is reduced from about 0.3 kWh to as low as 0.15 kWh. The major mechanism operating in this stage is that of absorption of organic material to the sludge floc. The reason for exclusion of primary sedimentation is to exploit the gut microorganisms which are present in raw sewage but removed during sedimentation. This feature is supposed to be essential for good first-stage operation and an independent A-stage clarifier means that these microorganisms are returned to the A-stage aeration basin.

The B stage completes oxidation of the sewage and it is loaded at 0.3 kg BOD/kg MLSS d. It operates aerobically and with a high solids retention time, consequently nitrification of ammonia can occur, with reported effluent ammonia concentrations of 5–10 mg/l. For a similar effluent quality, the overall loading for the A/B process can be increased by 50% compared to a conventional single-stage activated sludge plant.

4. The deep shaft process. This modification is a highly automated system in which the aerator comprises a concrete lined shaft which can be from 20 to 60 m deep. The shaft is divided into two independent process streams, a downcomer and a riser. Raw or settled sewage is introduced to the downcomer section, together with the return activated sludge. Air is injected into the downcomer section where it is

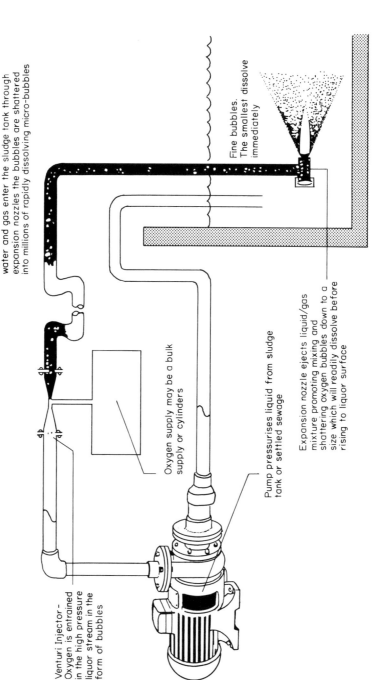

In this unit the bubbles are generated by introducing oxygen in fine bubble form into a high pressure stream of liquor. When the water and gas enter the sludge tank through expansion nozzles the bubbles are shattered into millions of rapidly dissolving micro-bubbles

Fine bubbles. The smallest dissolve immediately

Oxygen supply may be a bulk supply or cylinders

Pump pressurises liquid from sludge tank or settled sewage

Expansion nozzle ejects liquid/gas mixture promoting mixing and shattering oxygen bubbles down to a size which will readily dissolve before rising to liquor surface

Venturi Injector – Oxygen is entrained in the high pressure liquor stream in the form of bubbles

Figure 2.17 Schematic diagram of the VITOX high-pressure side-stream dissolver. (Diagram courtesy of Boc Ltd.)

carried down to the base of the shaft with the mixed liquor, and then up the riser section. The depth of the shaft results in both increased pressures down the shaft with a high degree of oxygen transfer and a longer contact time for the microorganisms with this dissolved oxygen. After leaving the shaft the mixed liquor passes to polishing aerators, which serve a dual function of removing residual BOD and also degassing the mixed liquor before it passes to conventional clarifiers.

Generally the rate at which sludge is returned to the downcomer is controlled automically and is related to the influent flow rate, thus producing a constant mixed liquor–solids concentration. The major benefits claimed for this process are that it occupies only a fraction of the land area that a conventional works would. In addition due to the shaft being sited underground, there is a much reduced visual impact at the site. In view of these advantages it is not surprising that 70% of the deep shaft systems currently in operation are sited in Japan. As of yet there is little reliable data to compare operating costs; however, initial reports suggest that there is a significant reduction in aeration costs.

5. The VITOX process. This process modification utilises pure oxygen to both mix the mixed liquor in the aeration basin and also provide a source of oxygen for bio-oxidation. Very efficient mixing is achieved by injecting gaseous oxygen into the throat of a venturi, through which the mixed liquor from the reactor is recirculated. This causes the formation of fine gas bubbles in the mixed liquor, which is then piped to the base of the reactor. Here it discharges into a 'spargerbar', which is a manifold of high-shear expansion nozzles which produce up to a 90% dissolution of oxygen. Consequently a series of high-pressure jets of very small bubbles of mixed liquor and oxygen are produced, resulting in the complete mixing of the reactor contents (Figure 2.17). Automatic oxygen saving is incorporated into the system by provision of oxygen probes which regulate the amount of oxygen injected, between preset minimum and maximum limits.

The VITOX system was originally intended as a method of uprating the oxygenation capacity of overloaded sewage treatment plants. In addition it has proved very efficient at the treatment of high-strength industrial wastes. Over 50 units have been installed in the UK and many operators report that VITOX units operate with very low sludge yields, this phenomena has been described as 'sludge burn-off', and the exact mechanism of this is unknown. More recently the potential of VITOX units as the first stage in a two-stage activated sludge process has been realised.

3 Process Fundamentals

3.1 REACTION KINETICS

Introduction

The biological reactions which take place in wastewater reactors are complex, and in order to provide mathematical descriptions of their effects many simplifying assumptions must be made. These assumptions will make the resulting model less exact, and it may not describe all aspects of the mechanisms under consideration. However, if these assumptions are clearly defined, it should ensure that the models are not applied inappropriately. In the construction of models of such phenomena as BOD removal or pathogen die-off it is usual to measure the phenomena under investigation, and vary in a known manner, those parameters thought to influence it. Appropriate mathematical expressions may then be fitted to the resulting data. The major simplifying assumptions used in models of biological growth processes which are applied to wastewater treatment systems fall into three categories.

The microbial population of a wastewater reactor comprises a large number of individual organisms of different species which vary physiologically and genetically. In view of the large number of organisms in such a population, the properties of the population may be described in terms of an 'individual' whose behaviour represents an average of the distribution of the behaviour of the population. Consequently the actual distribution of states is ignored. In addition to a range of states, microbial populations are also segregated into functionally discrete units or cells. The size of a population is usually measured by means of some easily determined parameter such as dry weight or nitrogen content. Thus several populations, comprising widely differing numbers of individuals, may have the same dry weight owing to differences in the size of the individuals which make up the population. Again most models make the assumption that the segregation of populations into individuals can be ignored, and models view a mixed population of microorganisms as biomass distributed continuously through the reactor. Finally, microbial populations undergo random deviations around mean values for all their growth processes; however, it is not possible to predict the behaviour of an individual microorganism. When the total number of cells involved is small these stochastic events may be significant. Such situations might include the phenomena of disinfection of treated effluents, or pathogen die-off in maturation ponds. However, for most phenomena this stochastic element is neglected, as random deviations from a large population will average out. Consequently, most models treat biological phenomena as deterministic processes.

Rate of reaction

The reactions which occur in a wastewater treatment process may be considered as a change in concentration of a substance or organism. These changes result from physical, chemical and biological processes and may often be modelled using simple reaction rate theory. This defines the rate at which a material or organism appears or disappears as the rate of reaction.

$$\text{A} \quad \rightarrow \quad \text{P} \tag{3.1}$$
$$\text{Reactant} \quad \text{Product}$$

Three different types of reaction are recognised namely, zero, first and second order and these all have the general equation:

$$\text{Rate} = k[A]^n \tag{3.2}$$

where n is the reaction order number, $[A]$ the concentration of number of reactants and k the reaction rate constant.

Zero-order reaction,

This proceeds at a rate which is independent of the concentration of starting product (A), and thus the rate of disappearance of A is given by

$$\frac{dA}{dt} = k[A]^0 \tag{3.3}$$

If the concentration of A at any time t, is given by C:

$$-\frac{dC}{dt} = k \tag{3.4}$$

thus at $t = 0$, $C = C_0$ and equation (3.4) may be integrated to give

$$C - C_0 = -kt \tag{3.5}$$

where k is the zero-order rate constant of units M/Vt.

Reactions which show zero-order kinetics will yield a straight line in a plot of the concentration of material or organism remaining at any time against time. This will have a slope of $-k$. A typical zero-order reaction is the oxidation of ammonia to nitrate by nitrifying bacteria (Chapter 8).

First-order reaction

The rate at which a first-order reaction proceeds is proportional to the initial concentration of reactant and is written:

$$-\frac{dC}{dt} = k[C] \qquad (3.6)$$

This may be integrated to

$$C = C_0 \exp(-kT) \qquad (3.7)$$

where C_0 is the value of C at $t = 0$. Inspection of equation (3.7) shows that a plot of $\ln C$ vs t will be linear with a slope of $-k$. This is the first-order rate constant which has units of t^{-1}. The majority of reactions of interest in wastewater treatment follow first-order kinetics—these include BOD removal from trickling filters and activated sludge reactors, and pathogen die-off in waste stabilisation ponds.

Second-order reaction

In a second-order reaction the rate is proportional to the second power of a single reactant:

$$-\frac{dC}{dt} = k[C]^2 \qquad (3.8)$$

this may be integrated to yield

$$\frac{1}{C} - \frac{1}{C_0} = kt \qquad (3.9)$$

Hence a plot of $1/C$ vs t will be linear with a slope equal to the second-order rate constant k, with units V/Mt. Second-order reactions are rare, and are generally observed in the oxidation of complex industrial wastes.

Effects of temperature on reaction rates

The reaction rate constant, k, is a lumped parameter incorporating many of the factors which affect reaction rate. Biological reactions in particular are influenced by a large number of environmental factors such as pH, oxygen concentration and light intensity. In later chapters, as our understanding of biological mechanisms increases, the simple models described above will be expanded to accommodate the effects of these parameters. One particularly important parameter which affects both chemical and biochemical reactions in a predictable way is temperature. Van't Hoff observed that as a rule of thumb reaction rates double for a 10 °C temperature rise. The effects of temperature on reaction rate may be described by the following equation:

$$\frac{d(\ln k)}{dt} = \frac{E}{RT^2} \tag{3.10}$$

where k is the reaction rate constant (t^{-1}), E the activation energy constant (J), R the gas constant (J/°C) and T the absolute temperature.

This may be integrated between the limits T_0 and T giving

$$\ln\frac{k}{k_0} = \frac{E(T - T_0)}{RTT_0} \tag{3.11}$$

where k and k_0 are the rate constants at T and T_0. In wastewater treatment systems the temperature change is over a relatively small range and thus E/RTT_0 can be considered as a constant. Equation (3.11) can therefore be approximated by

$$k = k_0\theta^{(T - T_0)} \tag{3.12}$$

where θ is a temperature coefficient characteristics of a particular wastewater process. Equation (3.12) is used widely in wastewater systems to model the effects of temperature on reaction rate. It should not, however, be used outside the range 5–25 °C as significant changes in the composition of microbial populations occur outside this range which invalidate the relationship.

3.2 HYDRAULIC CHARACTERISTICS OF WASTEWATER REACTORS

Introduction

The mixing characteristics of a reactor, and the manner in which a wastewater is introduced into the reactor, exert a considerable influence on the efficiency of treatment. Many important parameters are influenced by these hydraulic flow characteristics including BOD removal and settling properties in the activated sludge process, and BOD removal and pathogen removal in waste stabilisation ponds. In addition, mixing is one of the major factors contributing to discrepancies in results observed in the scale-up from laboratory to full-size plants. It is necessary therefore to be able to measure the mixing characteristics of a reactor, correctly interpret the results thus obtained and predict the likely behaviour of the reactor under different operating conditions.

Measurement of reactor mixing characteristics

The two extremes of reactor flow patterns are described as either plug flow or completely mixed. In a plug-flow reactor, the influent is envisaged as passing along the length of the reactor without experiencing any longitudinal mixing. As there is no intermixing, every element in the reactor experiences treatment for the same amount of time, namely the theoretical retention time. In a completely mixed reactor the influent is instantaneously mixed with the entire contents of the reactor and the contents of the reactor are homogeneous at every point. As a result of this the composition of the effluent is identical to the tank contents. In addition each particle will have a different retention time, with some leaving before the mean hydraulic retention time and others remaining longer.

As mixing characteristics are manifested by changes in the flow patterns of the wastewater within the reactor, this property is exploited in order to quantify these characteristics by construction of an age distribution function. Such an approach involves finding the time that individual elements of fluid remain in the reactor. Deviations of these elements from an average exit time are quantified by means of a dispersion coefficient. Age distribution functions are achieved by the introduction of a tracer to the reactor inlet and measuring its concentration at the outlet as a function of time. The tracer may be introduced as a discrete concentrated slug, known as an impulse function, which ideally should have a large magnitude and be of small duration. Alternatively it may be fed continuously at a lower concentration when it is known as a step signal.

The properties of an ideal tracer are:

1. It must be possible to detect easily at very low concentrations in the reactor outlet.

2. It should not be biodegraded within the reactor.

3. It must not react in any way with the reactor contents, for instance by binding to particulate material within the reactor.

4. It must be non-toxic.

5. It must not be visible in the environment.

Tracers which have been employed for this purpose include fluorescent dyes, radioisotopes and bacteriophage. No one tracer satisfies all these requirements and the advantages and limitations of the more popular ones are given in Table 3.1.

The output tracer response curves which result from impulse and step function inputs are illustrated in Figure 3.1 for idealised flow patterns of complete mixing and plug flow. In practice it is impossible to build a reactor in which the mixing is instantaneous, or in which no longitudinal mixing takes place. Mixing characteristics are found to vary between the two types and the terms arbitrary or dispersed flow are used to describe any degree of partial mixing between plug and completely mixed flow. Tracer outputs for typical dispersed-flow reactors are shown in Figure 3.2.

Tracer	Detection limit	Cost	Detection method	Half-life in environment
Rhodamine WT (moderate binding to particulate material)	0.01 µg/l	Moderate	Simple	High
Serratia marcescens phage (easy to use, but dies very quickly in some environments)	100 particles/l	Low	Simple	6 h–200 d
Bromophenol blue (low sensitivity and high cost make it expensive. Only slight binding, a good tracer)	20 µg/l	High	Simple	High
$^{82}Br^+$ (sophisticated apparatus required/a specialised technique)		High	Complex	36 h
Bacillus globigii spores (easy to use and detect. Once used it can persist in the environment giving a high background count)	10 spores/l	Moderate	Simple	9–15 months

Table 3.1 Properties of the major tracers of importance in tracing water movement.

At wastewater treatment plants, efficient hydraulic utilisation of reactors is frequently impaired by two factors: short-circuiting where a current of influent sewage exits from the tank very rapidly in considerably less than the theoretical retention time, and dead or stagnant zones which are generally in the corners of reactors and not significantly involved in the mixing process. For such phenomena as pathogen removal in which influent concentration is very high, a small degree of short-circuiting will cause a large decrease in removal efficiency. Conversely, the effect of dead zones is to reduce the effective volume of the reactor available for treatment. The presence of short-

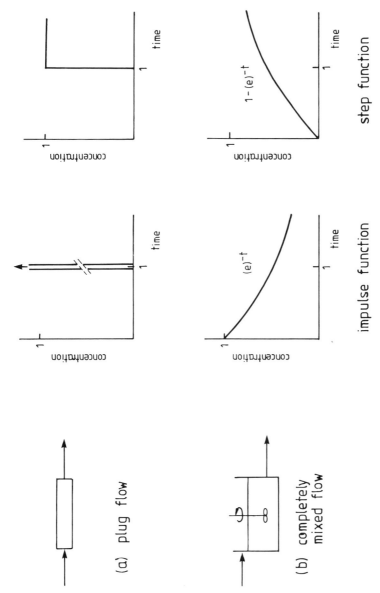

Figure 3.1 Response of completely mixed and plug-flow reactors to impulse and step function inputs of tracer.

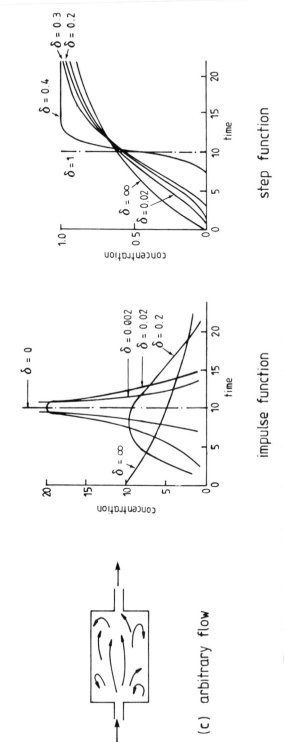

Figure 3.2 Response of dispersed or arbitrary flow reactor to impulse and step function inputs of tracer.

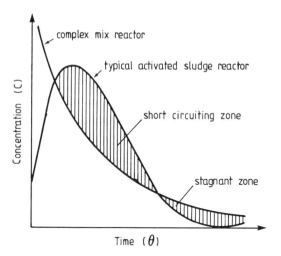

Figure 3.3 Areas of short-circuiting and incomplete mixing, revealed from reactor tracer studies.

circuiting and dead zones are revealed from the results of tracer studies (Figure 3.3).

3.3 MODELLING OF DISPERSION IN REACTORS

Several complex models are available to describe deviations from ideal flow, of which two have found a widespread application. These are the dispersion model and the tanks in-series model. Both models yield similar results if the mixing does not deviate far from plug flow. Although the tanks in-series model is simple and may be extended to a range of reactor types, the dispersion model is more widely used in practice. It is often stated, however, that the dispersion model is a better representation of small deviations from plug flow, whereas the tanks in series better describes small deviations from complete mixing.

Dispersion model

This model assumes that a pulse of tracer added to a reactor will start deviating from plug flow due to a number of factors (Figure 3.4). These factors include the presence of eddies in the reactor which hold the tracer back, turbulent mixing and the reactor velocity profile. This spreading or dispersion is represented by a dispersion coefficient (D) with units of m²/s. When $D = 0$, plug-flow conditions prevail and an increase in D represents an increase in the spreading. A value for D is obtained from the shape of the tracer exit curve. Two important parameters help to define the curve, namely T, which is the mean time of passage of the tracer, and the variance (σ^2) which is the square of the time taken for the tracer to leave the reactor and is a measure of the spread in time. As the spreading process is analogous to that of molecular diffusion it can thus be described by Fick's law:

$$\frac{\mathrm{d}C_t}{\mathrm{d}t} = D_m \frac{\mathrm{d}^2 C_t}{\mathrm{d}X^2} \qquad (3.13)$$

where $\mathrm{d}C_t/\mathrm{d}t$ is the change in fluid concentration with time, D_m the longitudinal or axial dispersion coefficient and $\mathrm{d}^2 C_t/\mathrm{d}X^2$ the rate of change of concentration gradient with distance.

If, however, the velocity of fluid through the vessel is constant, then the concentration at any point along the vessel will be a function of the liquid velocity in addition to the dispersion. Thus from equation (3.13) the concentration of fluid at any point X, along the vessel is given by

$$U\frac{\mathrm{d}C_t}{\mathrm{d}X} + \frac{\mathrm{d}C_t}{\mathrm{d}t} = D\frac{\mathrm{d}^2 C_t}{\mathrm{d}X^2} \qquad (3.14)$$

where U is the liquid velocity through a vessel of volume V, length L and flow Q. Thus,

$$U = \frac{QL}{V} \qquad (3.15)$$

Equation (3.14) may be written in dimensionless form using dimensionless forms of concentration ($C = C_t/C_0$), time ($\theta = t/\bar{t}$) and length ($z = X/L$) to become

$$\frac{D}{UL}\frac{\mathrm{d}^2 C_t}{\mathrm{d}z^2} - \frac{\mathrm{d}C}{\mathrm{d}z} - \frac{\mathrm{d}C}{\mathrm{d}\theta} = 0 \qquad (3.16)$$

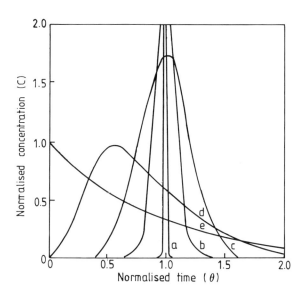

Figure 3.4 Typical tracer outputs for reactors which deviate from ideal plug flow: (a) $n = 50$, (b) $n = 30$, (c) $n = 10$, (d) $n = 4$, (e) $n = 1$.

The solution to equation (3.16) may be expressed graphically as a number of 'C'-curves for different values of D/UL where D/UL is referred to as the dispersion number (δ). These C-curves are continuous distributions whose spread increases as the dispersion increases. This spread may be measured as the variance of the distribution about its mean, and a measure of the variance of an experimentally derived C-curve offers a convenient way of obtaining the dispersion number.

Tanks in-series model

This is a simple technique to account for deviations in the behaviour of flow. It is based on the principle that if a number of completely mixed reactors are connected in series, their behaviour deviates from completely mixed towards plug flow, plug-flow conditions being approached as the number of tanks in series approaches infinity. Thus the mixing regime of any reactor can be described by the number of reactors in series for which its flow resembles. The tracer output curve which would result from N tanks in series may be determined from a mass balance of tracer concentration around the Nth tank (Figure 3.5).

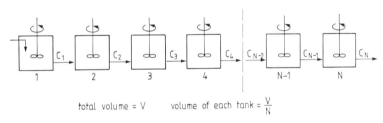

total volume = V   volume of each tank = $\frac{V}{N}$

Figure 3.5 Flow diagram of N completely mixed reactors arranged with their flow in series. (From Tomlinson and Chambers, 1979. Reproduced by permission of the Water Research Centre.)

Assuming that the tracer is inert and undergoes no reaction in the tank:

rate of change of = Amount of tracer − Amount of tracer
tracer in tank N entering   leaving

Thus for tank N of volume V:

$$\frac{dC_N}{dt}\frac{V}{N} = C_{N-1}Q - C_N Q \qquad (3.17)$$

This equation has a general solution for N tanks in series of

$$C = \frac{C_t}{C_0} = \frac{N^N}{(N-1)!} \cdot \theta^{N-1} \cdot e^{-N\theta} \qquad (3.18)$$

If theoretical curves are plotted of equation (3.18) for different values of N then a similar set of C-curves to those observed for the dispersion model are obtained (Figure 3.4).

3.4 CALCULATION OF DISPERSION NUMBERS

The solutions to the equations resulting from the tanks in series and dispersion model (3.16–3.18) are best evaluated graphically from tracer studies. In order to represent the results of tracer studies graphically in a form which they can be compared directly, the data are 'normalised' and plotted in dimensionless form. Effluent tracer concentration is plotted as relative effluent concentration (C_t/C_0) where C_t is tracer concentration in the effluent and C_0 is the initial concentration of tracer. Here C_0 is found from the area under the curve C_t vs t where

$$C_0 = C_t dT \qquad (3.19)$$

$$= \frac{1}{t_r} C_t dt \qquad (3.20)$$

If the tracer concentrations were measured at equidistant time intervals then equation (3.20) is approximated by

$$C = C_t dt \qquad (3.21)$$

The tracer retention time is plotted as relative retention time (t/t_r) where t is the time of measurement, and t_r is the theoretical mean retention time. Also, t_r may be found from a plot of C_t vs t which is a continuous function with the mean hydraulic retention time given by

$$t_r = \frac{C_t t}{C_t} \qquad (3.22)$$

It is now possible to construct a normalised residence time distribution of C vs T (Figure 3.4). Where this distribution is plotted with equidistant time increments, then its variance is described by the equation:

$$\sigma^2 = \frac{C_t T^2}{C_t} - 1 \qquad (3.23)$$

From the value of the variance calculated from the above equation, a measure of the deviation from ideal flow may be determined. The exact method for its enumeration depends upon the model being considered.

Dispersion model

The relationship between the dispersion number (D/UL) and the variance of a normalised C-curve is given by the equation:

$$\sigma^2 = \frac{2D}{UL} - 2\left(\frac{D}{UL}\right)^2 [1 - \exp(-UL/D)] \qquad (3.24)$$

Knowing a value for variance does not permit this equation to be solved analytically. If the dispersion number is low, however, this expression simplifies to

$$\sigma^2 \sim \frac{2D}{UL} = 2\delta \qquad (3.25)$$

A value for D/UL is now easily enumerated by iteration.

Tanks in series

The number of tanks in series which a reactor flow pattern approximates is easily found if a value for the variance is found from the appropriate C-curve. The variance of equation (3.23) being described by the expression:

$$\sigma^2 = \frac{1}{N} \qquad (3.26)$$

3.5 PRACTICAL PROBLEMS IN THE DETERMINATION OF DISPERSION NUMBERS

An essential feature of the activated sludge process is the return of a portion of the settled sewage from the secondary clarifier to the main reactor. Hydraulically this phenomena causes an increase in the wastewater flow, and in the presence of a tracer will appear to alter the mixing characteristics of the reactor. In the absence of recycle the added tracer would slowly wash out of the reactor and be lost from the system. When recycle operates, however, a steadily declining portion of the tracer makes a large number of repeated passages leading to the problem of multiple detection of a portion of the tracer. Consequently the calculated sum of tracer particles leaving the reactor is greater than the total amount added to the system, and an alteration in the form of the tracer response curve results. In a completely mixed reactor the return sludge would make no difference to the tracer response curve. As the effluent tracer concentration is

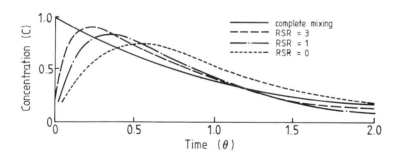

Figure 3.6 The effects of recycle ratio on the response of a plug-flow reactor to an impulse function input of tracer.

identical to the contents, the effluent concentration is unaffected by the reintroduction of a portion of the effluent. For other flow patterns, however, the response curve varies as a function of the number of tanks in series to which the flow pattern approximates. This effect is illustrated for a range of recycle ratios in Figure 3.6. It is apparent that as the flow pattern approximates plug flow, and for an increase in the recycle ratio, the tracer response curve approaches that of a completely mixed tank.

If the treatment plant layout permits, this effect may be overcome by using return sludge from a reactor which has not been pulsed with tracer, the sludge from the treated reactor being returned to an adjacent reactor. If this is not possible then the amount of tracer in the return sludge must be evaluated and appropriate corrections made.

3.6 EFFECTS OF REACTOR HYDRAULIC MIXING CHARACTERISTICS ON BOD REMOVAL

Plug-flow reactors

The influent to a plug-flow reactor passes through as a plug which experiences no loss or addition of BOD. The reactions which occur within each plug can therefore be considered as a batch process. Assuming that BOD removal follows first-order kinetics it can be described by equation (3.7) where t is the hydraulic retention time (V/Q). Using the notation for BOD this equation is written:

$$L_e = L_0 e^{-kt} \tag{3.27}$$

Completely mixed reactors

Due to the instantaneous mixing which occurs in a completely mixed reactor, the effluent is identical to the reactor contents. Removal of BOD is best described by the construction of a mass-balance equation.

Rate of change = Rate of inflow − (Rate of outflow

of BOD   + Rate of removal)   (3.28)

Thus for a reactor of volume V and assuming BOD removal is first order, then:

$$\frac{dL}{dt} V = QL_i - QL_e - kL_e V \tag{3.29}$$

Dividing equation (3.29) by V and assuming a steady state ($dL/dt = 0$), then

$$L_i = L_e + \frac{kL_e v}{Q} \tag{3.30}$$

The term V/Q can be replaced by the hydraulic retention time, t, and the equation rearranged to give

$$\frac{L_e}{L_i} = \frac{1}{1 + kt} \tag{3.31}$$

Dispersed flow

Mathematical modelling of dispersed-flow reactors is complex. Also it is difficult during the design stages to predict what the exact mixing characteristics of the reactor will be. Consequently, in practice, design of reactors is achieved by assuming either completely mixed or plug-flow conditions. An equation has been derived which considers dispersed-flow reactors as plug flow with a certain degree of dispersion. It therefore incorporates the dispersion number (d) and is known as the Wehner–Wilhelm equation:

$$\frac{L_e}{L_i} = \frac{4a\exp(\tfrac{1}{2}d)}{(1 + a)^2 \exp(a/2d) - (1 - a)^2 \exp(-a/2d)} \tag{3.32}$$

where:

$$a = \sqrt{(1 + 4ktd)} \tag{3.33}$$

The equation may be simplified by neglecting the second term in the

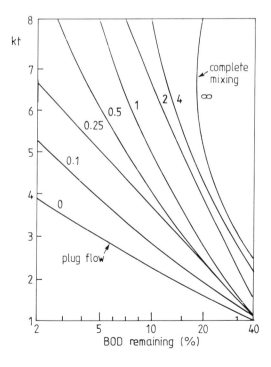

Figure 3.7 Thirumurthi chart for the Wehner–Wilhelm equation.

denominator which is small. This gives

$$\frac{L_e}{L_i} = \frac{4a\,e^a}{(1+a)^2}.$$

(3.34)

Despite this simplification, the Wehner–Wilhelm equation is rarely used directly. However, a design chart has been prepared from this equation known as a Thirumurthi chart which is useful in plant operation (Figure 3.7). For an existing reactor of a known dispersion number and known BOD removal coefficient, this chart allows an operator to select the hydraulic retention time necessary to give a required effluent BOD. Unfortunately Thirumurthi charts are rarely, if ever, used in practice. Inspection of this chart also shows that a greater degree of BOD removal is achieved in a plug-flow reactor as compared to a completely mixed reactor of the same volume and same BOD removal coefficient. Thus plants which suffer from overloading can frequently alleviate the problem by incorporating a degree of plug flow in the reactors, for instance by the use of baffles.

3.7 AERATION

Introduction

Aeration is an essential part of any biological wastewater treatment system, and where this is achieved mechanically it is usually the major energy-consuming process. With the increased emphasis on energy conservation, it is important to select the most efficient and econom-ical means of transferring oxygen into biological oxidation systems. In order to achieve this objective, and also to ensure that the treatment process functions effectively, an understanding of the basic principles of oxygen transfer are necessary.

Oxygen transfer mechanisms

The most widely used and accepted theory to describe the absorption of a gas by a liquid is known as the two-film theory. This envisages that oxygen transfer involves physical mass transport across a two-film layer that consists of a gas film (air) and a liquid film (the wastewater). Oxygen molecules are initially transferred across the gas film of the air bubble to the liquid film, and then by diffusion and convection across the liquid film into the bulk liquid. For a gas such as oxygen which has a limited solubility, the diffusion through the liquid film is slow and is the rate-limiting step. Consequently the rapid mass transfer through the gas film can be neglected. In wastewater treatment systems the oxygen then diffuses across the microbial cell

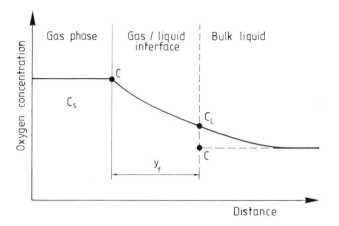

Figure 3.8 Schematic representation of the two-film theory of oxygen transfer between a gaseous and a liquid phase.

wall and is utilised in cell metabolism. As the rate of metabolism in wastewater treatment systems is limited by the availability of BOD, the rate of microbial metabolism will determine the rate of oxygen demand. If this oxygen demand exceeds the rate at which oxygen diffuses through the liquid film, the oxygen will be limiting. It is the role of an aerator, therefore, to increase the rate of oxygen transfer from the liquid film to the bulk liquid at a rate sufficient to meet the oxygen demands of metabolism.

If the diffusion of gas across the liquid film is considered as a mass transfer, then the two-film theory may be modelled mathematically by the application of Fick's law. For an ideal situation in which a stationary liquid is in contact with the air, at the interface between the two phases there will be a liquid film in which the dissolved oxygen molecules are concentrated (Figure 3.8). The driving force for the transfer of oxygen from the gas phase to the interface is the partial pressure of oxygen in the gas phase, and the driving force at the interface is the oxygen concentration gradient between the interface layer and the bulk liquid. Fick's law states that gases move spontaneously from a region of high concentration towards a region of low concentration, and the greater the concentration gradient the greater the rate of transfer. Thus, for any transfer process, the rate of transfer is the product of a driving force and a transfer coefficient, which yields a partial differential equation in space and time:

$$\frac{dM}{dt} = -D_L A \frac{dC}{dy_f} \tag{3.35}$$

where dM/dt is the mass transfer rate (mass/time), D_L the diffusion coefficient (area/time), A the cross-sectional area across which oxygen is diffusing (area) and dC/dy_f = oxygen concentration gradient across an interface of thickness y_f (mass/volume length).

However, as the interface layer y_f is only a few molecules thick, the differential quantity can be replaced with a linear approximation of the oxygen concentration gradient:

$$\frac{dC}{dY_f} \sim \frac{C_s - C}{y_f} \qquad (3.36)$$

where C_s is the oxygen concentration in the interface layer and C the oxygen concentration in the bulk liquid. If this is substituted in equation (3.35) an ordinary differential equation in time is obtained:

$$\frac{dM}{dt} = -D_L A \frac{C_s - C}{y_f} \qquad (3.37)$$

If this is now divided by the volume of the liquid phase (V) then:

$$\frac{1}{V}\frac{dM}{dt} = -D_L \frac{A}{V} \frac{C_s - C}{y_f} \qquad (3.38)$$

The mass transfer term now has units of mass/volume time, in other words concentration/time. It can therefore be replaced by dC/dt, thus:

$$\frac{dC}{dt} = -D_L \frac{A}{V} \frac{C_s - C}{y_f} \qquad (3.39)$$

The interface y_f is only a few molecules thick, and consequently it is difficult to measure. To circumvent this, it is combined with the diffusion constant D_L to yield a new coefficient K_L, which is known as the gas transfer with units of length/time.

$$K_L = \frac{D_L}{y_f} \qquad (3.40)$$

This can now be replaced in equation (3.39) to give

$$\frac{dC}{dt} = -K_L \frac{A}{V}(C_s - C) \qquad (3.41)$$

Equation (3.40) shows that under quiescent conditions, the main factors which affect the gas transfer coefficient are the interface thickness and the rate of molecular diffusion. Under the turbulent conditions which exist in aerated reactors, however, the surface film layer is continually being renewed so that deoxygenated liquid replaces oxygen-saturated liquid at the interface from the bulk liquid. In addition the rate of diffusion is supplemented by the rate of surface film renewal and consequently the degree of reaeration is a function of the amount of turbulence. As the turbulence increases, finer and finer droplets are produced, with an increasing gas transfer rate. Increasing turbulence also causes an increase in the interfacial area A, and this cannot be measured in practice. Consequently a lumped parameter is produced by combining the quotient A/V (known as the specific surface and denoted as 'a'), and the gas transfer coefficient. This is known as the overall oxygen mass transfer coefficient $K_L a$ and has units of t^{-1}. Equation (3.41) now takes the form:

$$\frac{dC}{dt} = K_L a(C_s - C) \tag{3.42}$$

Under conditions of turbulence equation (3.42) may be integrated for two concentrations of oxygen in the bulk liquid: C_0 at zero-time (t_0) and C at any given time (t) to yield:

$$K_L a = \frac{2.303 \log[C_s - C_0/C_s - C]}{t - t_0} \tag{3.43}$$

This equation allows the overall mass transfer coefficient to be calculated either analytically or graphically as described below. The overall mass transfer coefficient provides a useful guide to the efficiency of an aerator and is an important design parameter.

Techniques for the evaluation of $K_L a$

Here $K_L a$ is a measure of saturated fluid turnover time and is a function of the aeration equipment, tank geometry, and wastewater characteristics. In order to standardise results, aeration equipment is evaluated at the manufacturers in standard test-tanks, under standard conditions, which leaves $K_L a$ simply a function of the aeration equipment. In Europe aerators are tested in tapwater at 10 °C, 760 mm Hg barometric pressure and zero dissolved oxygen concentration. These conditions are slightly different in the USA where the temperature is 20 °C, and in the UK a surfactant is added to alter the properties of the tapwater such that they more closely resemble wastewater. Two techniques are routinely used to evaluate aeration transfer efficiency.

1. The unsteady-state method involves the chemical deoxygenation of clean water, followed by aeration during which the rate of increase in dissolved oxygen with time is measured. The water is deoxygenated catalytically by the addition of sodium sulphite which reacts with oxygen according to the reaction:

$$2Na_2SO_3 + O_2 \rightarrow 2Na_2SO_4 \tag{3.44}$$

Stoichiometrically, 8 mg/l of sodium sulphite are required for each milligram of O_2/l, but it is conventional practice to add twice this amount. In addition, 0.05 mg/l of cobalt is also added in order to increase the rate of oxidation of any excess sulphite, which would otherwise interfere with the measurement of oxygen transfer rates. Knowing the saturated concentration of dissolved oxygen (C_s) at the temperature of the test allows $K_L a$ to be determined graphically from the slope of the line obtained from a plot of $\log(C_s - C)$ against time.

2. The steady-state test is performed in mixed liquor after the aerator

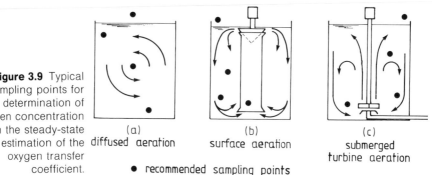

Figure 3.9 Typical sampling points for determination of oxygen concentration in the steady-state estimation of the oxygen transfer coefficient.

(a) diffused aeration

(b) surface aeration

(c) submerged turbine aeration

● recommended sampling points

has been installed. It is not a simple technique as dissolved oxygen must be monitored at several points in the aeration basin to obtain representative values–recommended sampling points for a number of aeration devices are shown in Figure 3.9. The aim is to ensure that each sampling point measures the oxygen concentration of an equal volume of aerated liquid. In addition, corrections must be made for the oxygen uptake rate of the sludge caused by microbial metabolism. The oxygen transfer is calculated from the equation:

$$\frac{dC}{dt} = K_L a(C_s - C_0) - R_0 \qquad (3.45)$$

Where R_0 is the oxygen uptake of the mixed liquor in mg Q_2/l. Since at steady state the term dC/dt is zero, then equation (3.45) becomes

$$K_L a = \frac{R_0}{C_s - C_0} \qquad (3.46)$$

This equation may be used analytically to calculate aerator performance using several sets of operating data.

Factors affecting the magnitude of $K_L a$

Manufacturers usually test and rate an aerator under standard conditions of temperature (20 °C), pressure (760 mm Hg) and in clean water. The oxygen transfer rates obtained under these conditions are known as the standard oxygen transfer rate (SOTR). However, many factors influence oxygen transfer mechanisms and these include wastewater composition, temperature, basin geometry, turbulence and prevailing pressure. These factors make each field application of an aeration device unique, and it is unlikely that manufacturers' quoted transfer efficiencies will be realised. Field oxygen transfer rates are calculated from standard transfer rates by the use of correction factors. Three such factors are required to translate aerator performance under standard conditions to likely performance under field conditions.

The alpha factor is determined from the oxygen transfer coefficients of the aerator in tapwater and wastewater:

$$\alpha = \frac{K_L a \text{ wastewater}}{K_L a \text{ tapwater}} \qquad (3.47)$$

Determination of meaningful alpha factors is extremely difficult. Ideally alpha-factor testing should be performed using the actual wastewater to be aerated. For plants which have not yet been built, this may create problems, and influent wastewater is a poor substitute for mixed liquor, with lower alpha factors generally being recorded. The mixing regime of the aeration tank presents special problems in the determination of alpha factors if it is either plug-flow or tapered aeration. Changes in the composition of the wastewater as it passes along the tank results in alpha factors which are as low as 0.3 at the tank influent, but up to 0.8 at the tank effluent. Current UK practice is to add 5 mg/l of synthetic anionic detergent to tapwater under standard conditions. This is meant to provide aeration rates which approximate those observed under field conditions more closely and results in measured alpha factors being closer to unity and consequently less dependence is placed upon them, resulting in a reduction of errors in aerator selection.

The beta factor is defined as the ratio of the saturated dissolved oxygen concentration in wastewater and tapwater.

$$\beta = \frac{C_s \text{ wastewater}}{C_s \text{ tapwater}} \qquad (3.48)$$

It is determined by aerating identical volumes of tapwater and wastewater to saturation and accurately measuring the saturated dissolved oxygen concentration. The beta factor is affected by a large number of variables and process conditions, such as solids concentration, dissolved organic material and temperature. Wastewaters have a sufficiently high concentration of dissolved solids to reduce the saturated dissolved oxygen concentration significantly; however, beta factors do not vary over such a broad range as alpha factors. The major problems in the determination of beta factors are technical ones resulting from interferences in the chemical determination of dissolved oxygen concentrations. Since beta factors also vary with wastewater quality, several tests must be performed in order to establish a representative value.

The theta factor is an empirical term which corrects for the effects of temperature on oxygen transfer. Many factors such as liquid viscosity, surface tension and oxygen diffusivity all affect oxygen transfer and are in turn affected by changes in temperature. The theta factor is a lumped parameter which attempts to correct for all these factors and this has led to a variety of correction factors, with both geometric and arithmetic relationships proposed. The consensus of opinion suggests that a geometric model such as equation (3.12), with a theta factor of 1.024 proves most appropriate.

$$K_La(T) = K_La(20°)\theta^{(T-20)} \tag{3.49}$$

However, due to inherent inaccuracies present in such approximations, it is recommended that temperature corrections in excess of 10 °C are avoided if at all possible.

With a knowledge of the appropriate correction factors, field oxygen transfer rates (OTR) may be calculated from the manufacturer's standard oxygen transfer rate (SOTR) according to the equation:

$$OTR\,(kg\,O_2/kWh) = \alpha\left(\frac{\beta C_s - C}{C_{20}}\right)1.024^{(T-20)}SOTR \tag{3.50}$$

where C_s is the saturated oxygen concentration of tapwater at the prevailing pressure, C the required operating dissolved oxygen concentration and C_{20} the saturated oxygen concentration of tapwater at 20 °C.

When comparing the oxygenation capacity of different aeration devices, a commonly used criteria is the oxygenation capacity (OC) which has units of $kg\,O_2/h$. This is defined as the rate of absorption of oxygen during aeration of completely deoxygenated liquor at a given temperature, thus:

$$OC = K_LaVC_s \tag{3.51}$$

where V is the volume of liquor under aeration and C_s the saturation concentration of dissolved oxygen at the test temperature.

If the oxygenation capacity of an aerator is then divided by the amount of power supplied to the aeration system, a figure for the oxygenation efficiency (OE) is obtained which has units of $kg\,O_2/kWh$.

3.8 SEDIMENTATION THEORY

Introduction

The process of sedimentation which occurs in a sedimentation tank may be defined as the removal of solid particles from a suspension by settling under gravity. In wastewater treatment, particle sedimentation is exploited in grit removal, primary sedimentation and secondary sedimentation. A wide range of solids are encountered in a wastewater which exhibit a range of particle sizes and densities. Some of these particles will not change their properties during sedimentation (discrete particles), whereas others will agglomerate and flocculate, therefore undergoing large changes in their settling properties (flocculent particles).

The settlement of discrete particles has traditionally been described using Stokes's law. This quantifies the factors which affect the settling velocity of a small solid particle in a quiescent liquid and is given by

$$V = \frac{2g}{9}\cdot\frac{r^2}{n}\cdot(d_2 - d_1) \qquad (3.52)$$

where V is the settling velocity, r the particle radius, g the gravitational acceleration, n the kinematic viscosity of suspending fluid, d_1 the density of suspending fluid and d_2 the density of particle.

It is not possible, however, to apply this equation directly to the process of wastewater sedimentation since neither particle diameter nor density is known. In addition, since the particles are irregular, it is not practical to determine these properties. A further complication is that the particles encountered in a wastewater tend to agglomerate and form flocculant suspensions. As the liquid depth increases, the likelihood of contact between particles also increases. This results in the formation of larger and heavier particles with increased settling velocities. The differences in settlement exhibited by discrete and flocculent particles are shown in Figure 3.10. What Stokes's law does illustrate, however, is that settling velocity is influenced by particle diameter and density as well as the properties of the suspending fluid. Such information proves useful in understanding the settlement of particles.

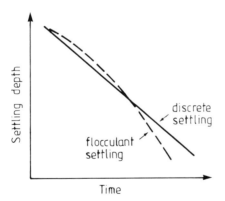

Figure 3.10 Settling velocities exhibited by discrete and flocculent particles.

For the purpose of sedimentation tank design, four distinct modes of settling have been described which are a function of particle size and particle interaction (Figure 3.11). These settling modes may occur independently during a wastewater sedimentation process (e.g. type 1 settling during grit removal), but more usually, more than one type occurs simultaneously.

Type I (discrete) settling. The unhindered settling of discrete particles such as sand, as predicted by Stokes's law, i.e. the particle

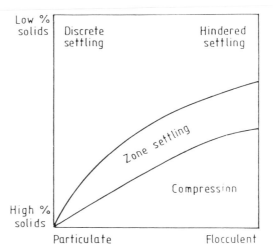

Figure 3.11 Relationship between settling mode and solids concentrations.

undergoes no change in shape, size or density during settlement. This type of sedimentation predominates in grit-removal chambers.

Type II (flocculant) settling. The settlement of colloidal and larger particles in dilute suspension. These particles demonstrate agglomeration and flocculation, thus the settling velocity increases with depth.

Type III (zone or hindered) settling. The concentration of flocculent particles is sufficiently high to allow interparticle forces to bind them together in a lattice structure. Consequently the particles no longer settle independently, but as a mass or unit with a distinct solid/liquid interface visible. The liquid which is displaced by the settling particles will result in a reduced settling velocity, and this is known as 'hindered settling'. The rate at which settling occurs with such particles is controlled by the rate at which water passes upwards through the mass. A settling column test on such a suspension will exhibit distinct zones of solids at different concentrations and hence the name of zone settling.

Type IV (compressive) settling. At the base of a sedimentation tank, the concentration of solids is so high that the particles are actually in contact with each other. Thus each layer of particles is supported by the layer above it. Settling in this region is no longer governed solely by the size of the particle since further settling can only occur by the forcing out of water from the compressing particles and this requires adjustments in the matrix which forms the sludge blanket. Thus the settling rate is determined by the compressible properties of the sludge and settling in this region is slow.

Type I settling is assumed to occur in gravity grit chambers, type II settling occurs in the primary settlement of wastewaters and type III settling is the major mechanism of importance in clarifiers which follow the activated sludge process.

Primary sedimentation tanks

Primary sedimentation tanks receive an influent which has been subjected to preliminary treatment. Consequently the majority of particles which demonstrate class I settlement have been removed in the grit channels. Design is based therefore on the assumption of class II settlement properties. In view of the shortcomings in traditional models of particle settlement, analysis of experimentally determined settling velocities will provide the most reliable information to ensure rational design.

The necessary information is obtained from settling tests performed using a test cylinder filled with a uniformly mixed suspension of wastewater. The cylinder, which should be the same height as the proposed sedimentation tank, is allowed to settle under quiescent conditions (Figure 3.12), and samples removed at intervals from sampling ports fixed at different depths on the cylinder. The suspended solids concentration of these samples is determined and, in addition, the height to which the solids have settled in the cylinder. It is apparent from consideration of the mechanism of class II settlement, that the most important criteria for design is the provision of adequate detention time, in order to allow settlement to occur. The results obtained from settling analysis may be expressed in the form of a depth–time grid which allows calculation of the required depth and detention time necessary to achieve a given removal of settleable solids. These are constructed by determining the percentage solids

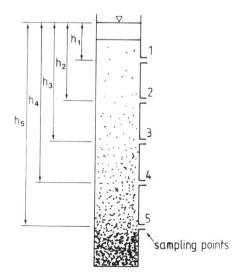

Figure 3.12 Laboratory settling column for the determination of primary sedimentation tank design parameters.

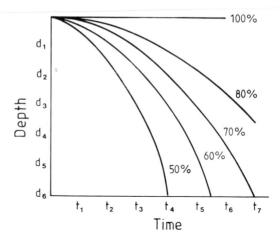

Figure 3.13 Isoconcentration curves drawn through points of equal percentage removal of solids, using data obtained from the laboratory settling column.

removal for each sample analysed and then plotting these as curves of equal percentage removal as illustrated in Figure 3.13.

Many people have shown, however, that the area of the tank is more important than its depth or volume, since the upward flow velocity is determined by the volumetric flow rate and the tank area. The required tank area is determined therefore from the surface loading rate, which is defined as the flow rate to the tank divided by its surface area (Q/A). The surface loading rate necessary to achieve a given percentage of solids removal is calculated by dividing the depth of tank selected by the time required to achieve the necessary removal at that depth. As the degree of flocculation of the suspended particles will vary with their concentration, then a series of tests will be required over the range of settleable solids concentration expected in the wastewater. This will result in a family of settling curves as depicted in Figure 3.13. It is usual to expect a settleable solids removal of 50–60% in primary sedimentation tanks.

The results of laboratory settling tests represent the maximal rates obtained under ideal conditions. Practical continuous-flow sedimentation tanks do not behave in exactly the same way since:

1. Uniform flow through the inlet and outlet zones is not achieved.

2. Quiescent conditions are not achieved due to the existence of turbulence.

3. The influent wastewater is of variable composition.

4. In hot climates density currents are produced by temperature differences within the tank.

For these reasons, tank performance will not meet that achieved in a settling column and safety factors are employed to compensate for deviations from ideal behaviour. It is usual to increase detention times by a factor of 1.5–2.0, and to reduce surface loadings by 1.25–1.75.

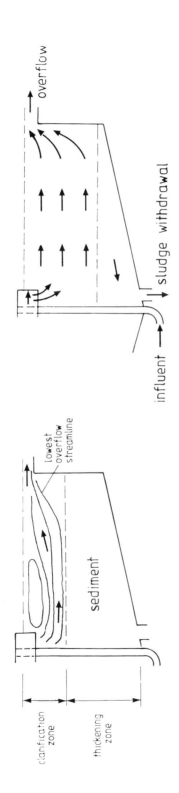

(a) TYPICAL PATTERN

(b) IDEAL PATTERN

Figure 3.14 Typical secondary sedimentation tank flow patterns.

In the absence of suitable laboratory settling data (which are often unavailable as samples of typical influent wastewater are not available), tanks are designed by selecting a suitable surface loading rate based on the performance of existing plants. It should be cautioned, however, that the performance of sedimentation tanks is very environment specific, as a result of differences in temperature and wastewater composition. Consequently, selected surface loading rates should be obtained from sedimentation tanks operating within the locality of the new plant, and not from quoted design data.

Secondary sedimentation tanks

Secondary sedimentation tanks should provide two functions: good clarification resulting in a low effluent solids concentration and a high thickening of the sludge underflow. Both criteria should be considered in design calculations and the more conservative design area selected. It is usually observed that the thickening component is the most critical, but despite this the concepts which relate to design for thickening functions are less familiar than those which relate to design for clarifications.

Clarification

In order to ensure adequate clarification, the tank area must be such that the hydraulic loading to the tank does not exceed the settling velocity of the slowest settling particles. This is generally taken to be the zone settling velocity of the sludge at its operating MLSS concentration and is determined from settling column tests. The theoretical hydraulic retention time (t) in the clarifier section of a tank with an effluent flow rate of Q_e is given by

$$t = \frac{hA}{Q_e} \tag{3.53}$$

where A is the area of clarifier section of tank and h the depth of clarifier section of tank.

Thus if a particle is to be removed during the period t, it must settle a maximum distance h. If the settling velocity of the slowest settling particle to be removed is v, then

$$vt = h \tag{3.54}$$

If equations (3.53) and (3.54) are combined, this yields:

$$v = \frac{Q_e}{A} \tag{3.55}$$

The term Q_e/A is known as the surface settling rate or overflow velocity. It is apparent from this equation that a particle will be removed if it settles as fast or faster than the vertical fluid velocity, and thus surface settling rate would appear to be a good parameter for design of settling tank clarification function. Equation (3.55) suggests that the degree of solids removal is independent of tank depth. Whereas this may be true for discrete particles, activated sludge particles demonstrate flocculation, and thus the deeper the tank then

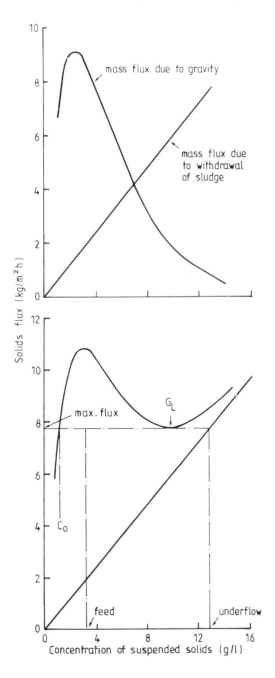

Figure 3.15 The components of the solids flux test, and the resulting total flux.

the greater the degree of flocculation and the greater the portion of solids removed. It is also assumed that ideal conditions exist in a sedimentation tank, in particular that the solids flux is vertical. This is of course not the case, with turbulence and short-circuiting occurring. A typical clarifier flow pattern is illustrated in Figure 3.14.

In addition, use of this parameter does not ensure that a final sedimentation tank will achieve a good solids thickening, since the overflow velocity is not related to the degree of solids concentration.

Thickening theory

Upon leaving the clarification zone of a sedimentation tank, the solids enter the thickening zone where their concentration is increased to the underflow value. In order to achieve a desired underflow solids concentration, sufficient area must be provided so that the solids are applied at a rate less than the rate at which they are able to reach the bottom of the tank. The thickening capacity of a tank is generally calculated from solids flux theory. This is based on the concept that there is a maximum quantity of solids which can be handled by a sedimentation tank at a given underflow removal rate without affecting performance. Solids are transported to the bottom by two velocity components. One of these is the subsidence due to gravitational forces (v), and the other results from a downward flux due to sludge withdrawal from the base of the tank (u). The solids flux which results for the gravitational component is termed a settling flux (G_s) and defined as the product of the solids concentration and settling velocity:

$$G_s = vX \qquad (3.56)$$

The solids flux due to solids removal is termed the bulk flux (G_b) and is the product of the solids concentration and the underflow withdrawal velocity (u).

$$G_b = uX \qquad (3.57)$$

At a solids removal rate of Q_u for a tank of area A, the underflow velocity is given by

$$u = \frac{Q_u}{A} \qquad (3.58)$$

The total rate at which solids of concentration X travel downwards in the final sedimentation tank is the sum of the two fluxes and is termed the solids flux (G), thus:

$$G = G_s + G_b \qquad (3.59)$$

Substituting from equations (3.56) and (3.57), this may also be written:

$$G = vX + \frac{Q_u}{A}X \qquad (3.60)$$

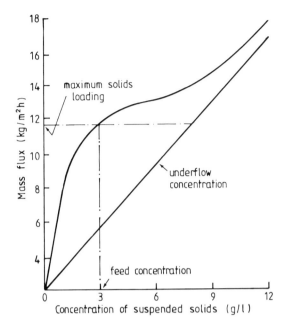

Figure 3.16 The total flux curve at the critical solids withdrawal rate.

The two components of the solids flux and the total flux are illustrated schematically in Figure 3.15, and this illustrates the effect that variations in the solids concentration will have. The exact form of this curve is heavily dependent upon the rate of sludge withdrawal. At typical withdrawal rates the curve has a maximum followed by a minimum, and the value of the flux at the minimum is termed the limiting flux (G_L). The limiting flux represents the maximum solids loading which may be applied to the tank in order to thicken a sludge from its influent concentration to a value above the solids concentration at the limiting flux. This parameter thus establishes the solids handling capacity of the tank. With increasing withdrawal rates, however, the maximum and minimum of Figure 3.15 will converge until, at a critical underflow rate, they merge to yield a point of inflection (Figure 3.16). Under these conditions the limiting flux is found from equation (3.60) where v_f and X_f become the values of the mixed liquor settling velocity and inlet solids concentration respectively, thus:

$$G_L = v_f X_f + \frac{Q_u}{A} X_f \qquad (3.61)$$

The actual solids loading applied to a sedimentation tank is given by

$$G(\text{appl}) = \frac{X_f(Q + Q_u)}{A} \qquad (3.62)$$

If the tank is not to be overloaded and lose solids over the weir, then

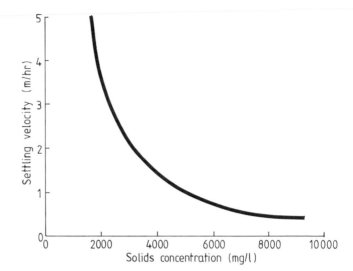

Figure 3.17 The relationship between sludge-settling velocity and solids concentration, which is used in the determination of the coefficients V_0 and k.

the solids loading applied to the tank must not exceed the maximum predicted loading, i.e.

$$G(\text{appl}) < G_L \tag{3.63}$$

Substituting from equations (3.61) and (3.62):

$$X_f\left(\frac{Q}{A} + \frac{Q_u}{A}\right) < X_f\left(v_f + \frac{Q_u}{A}\right) \tag{3.64}$$

This equation reduces to

$$\frac{Q}{A} < v_f \tag{3.65}$$

It will be seen that this equation is identical in form to equation (3.55) and thus design based on thickening theory also ensure that the clarification requirements of a tank will be met.

The settling flux component of the solids flux (G_s), is dependent upon the settling properties of the activated sludge and is thus largely determined by the operating conditions prevalent in the aeration basin. The bulk flux (G_b), however, is controlled directly by the plant operator, as the value of u is varied by varying the rate of thickened sludge removal. If it is assumed that all the solids fall to the base of the sedimentation tanks, then the total mass of solids (M) which require thickening per unit time is given by

$$M = Q_u X_u \tag{3.66}$$

where Q_u is the flow rate and X_u the solids concentration of the sludge removed from the tank. The downward velocity which results from this solids removal is given by equation (3.58) and substituting for Q_u

from equation (3.66):

$$u = \frac{M}{X_u A} \qquad (3.67)$$

It is apparent from this that for a given solids loading (M) to a sedimentation tank of area A, then the rate of sludge withdrawal (u) is inversely related to the concentration at which the solids are withdrawn. The required tank area is governed by the limiting solids handling capacity which as shown in Figure 3.15, is the minimum solids flux (G_L) given by

$$A = \frac{M}{G_L} \qquad (3.68)$$

This may also be expressed in terms of the influent solids loading as

$$A = \frac{Q_0 X_0}{G_L} \qquad (3.69)$$

where Q_0 is the influent and recycle flow rate to the sedimentation tank and X_0 is the influent solids concentration. Thus solids flux theory allows the graphical determination of G_L from settling tests and this allows the calculation of tank area for a given underflow solids concentration.

An alternative analytical approach to sedimentation tank design is based on the observation that the settling flux components of equation (3.60) are functions of the settling properties of the activated sludge and are dependent upon the solids concentration. This relationship may be determined from batch settling tests at different solids concentrations. An empirical expression for the settling rate has also been derived which takes the form:

$$V_s = V_0 e^{-kX} \qquad (3.70)$$

Where V_0 and k are constants which describe the sludge settling behaviour and are determined by regression analysis (Figure 3.17). If this expression is now substituted in equation (3.60):

$$G = \left(\frac{Q_u}{A}\right) X + V_0 e^{-kX} X \qquad (3.71)$$

The required tank area is governed by the limiting solids handling capacity which, as shown in Figure 3.15, is the minimum solids flux (G_L) which is the value of G at the minimum point on the curve of G vs X. As the rate of sludge withdrawal increases, the maximum and minimum on Figure 3.15 converge, until at a critical underflow rate ($[Q_u/A]$crit), they merge to a point of inflexion (Figure 3.16). Here G_L is found by differentiating equation (3.71) twice with respect to X:

$$\frac{dG}{dX} = \frac{Q_u}{A} + V_0 e^{-kX}(1 - kX) \qquad (3.72)$$

$$\frac{d^2G}{dX^2} = V_0 k e^{-kX}(kX - 2) \tag{3.73}$$

At the critical point $dG/dX = d^2G/dX^2 = 0$.
Thus from equation (3.73)

$$0 = V_0 k e^{-kX}(kX - 2) \tag{3.74}$$

and

$$kX = 2 \tag{3.75}$$

substituting $kX = 2$ and $dG/dX = 0$ into equation (3.72) gives:

$$0 = \frac{Q_u}{A} + V_0 e^{-2}(1 - 2) \tag{3.76}$$

$$\left(\frac{Q_u}{A}\right)_{crit} = V_0 e^{-2} \tag{3.77}$$

$$= 0.13 V_0 \tag{3.78}$$

Both equations (3.54) and (3.77) suggest that the theoretical depth of a sedimentation tank is unimportant. However, as activated sludge particles are flocculant, and therefore their settling velocity will change with depth, then some specification to aid in selecting depth would seem appropriate. The determination of the depth of a sedimentation tank is still largely empirical. Many sedimentation design criteria state that the hydraulic retention time of a tank should not be less than a stipulated value which generally ranges from 2 hours upwards. The problems with specifying retention times, however, is that the entire tank volume is used in its calculation when a large part of the tank is in fact occupied by sludge. A second problem is that the retention time is established solely on the basis of the waste flow rate and thus the thickening function of a tank is ignored. If the settling properties of activated sludge were predictable and of known good quality, then the effects of depth would be relatively unimportant. However, this is not the case and the effects of depth should be evaluated. This is achieved by constructing a series of batch flux curves derived from sedimentation tests performed at different initial depths. Various combinations of area and volume which produce an identical performance efficiency may then be evaluated. In choosing an appropriate depth, several other operational factors are important and require consideration. It is important that sufficient thickening depth be provided to maintain an adequate sludge blanket such that during recycle, clarified liquid is not withdrawn with the sludge. In addition, during operation the tank is likely to be subjected to periods of temporary overloading, for instance when other tanks at the plant are down for maintenance. Sufficient depth to store solids in the thickening portion of the tank must be available.

4 Microorganisms Exploited in Wastewater Treatment

4.1 INTRODUCTION

With the exception of certain industrial wastes, wastewaters provide an ideal growth medium for a large number of different microorganisms, and these microorganisms play a key role in all stages of biological wastewater treatment. No single organism is capable of utilising all of the wide variety of inorganic and organic compounds found in wastewaters. Consequently a diverse ecosystem will develop, feeding directly on the incoming sewage and preying on the organisms so produced. The exact composition of this community will depend on the outcome of competition for a limited and varied food supply, and this outcome will be influenced by environmental parameters such as pH and temperature. It is the aim of plant design and operation to create favourable conditions such that the desired microorganisms can proliferate, and effect the necessary treatment. Effective design therefore, requires a knowledge of the types of microorganisms which are required for each of the different treatment processes, and the environmental conditions under which they demonstrate their maximum growth potential. There has been a large increase recently in our knowledge of wastewater ecology, and this is providing engineers with valuable information on plant operation and control. It is appreciated that microbial taxonomy proves daunting for engineers who have little or no biological background. However, the time involved in mastering the terminology and recognising the differences between the various microorganisms is well spent. In the right hands a microscope may be used to diagnose rapidly a wide range of wastewater treatment plant operating problems.

Although living organisms were originally classified into two kingdoms, Plantae and Animalae, such a simple classification is no longer tenable. Microorganisms include certain groups which have plant-like properties (the green algae), some which are animal like (the protozoa) and some with characteristics of both kingdoms (the fungi). Consequently biologists now recognise a further kingdom, the Protista, which is a diverse group of phyla which do not fall naturally into either of the other kingdoms. Of the three kingdoms, it is the

· Protista which is of most interest to sanitary engineers, as it contains the bacteria, protozoa and algae, and these phyla are the most important microorganisms involved in wastewater purification.

With the advent of electron microscopy in the 1950s, a fundamental difference was noticed between the internal structure of the bacterial cell, and that of all other cell types. With the exception of bacteria, all cells were observed to contain complex and well-differentiated internal structures (known as organelles), surrounded by membranes. These membranes resemble the membrane which surrounds the cell itself. Typical organelles include the cell nucleus, mitochondria, ribosomes and Golgi apparatus. Within the actual cell nucleus, the

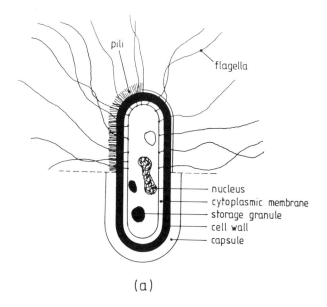

(a)

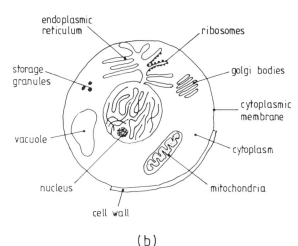

(b)

Figure 4.1 Typical cell structures: (a) prokaryotic cell, (b) eukaryotic cell.

cellular genetic material was organised into recognizable structures called chromosomes. These cells are known as eukaryotes (Figure 4.1).

By contrast bacterial cells contain only protoplasm, bounded by a semi-permeable cytoplasmic membrane. This is covered in turn by a porous cell wall. The cytoplasm contains the nuclear material, and this is not bounded by a discernible membrane separating it from the cytoplasm. In addition no internal membranes are found isolating recognisable structures. Organisms possessing these properties are known as procaryotes (Figure 4.1). Thus microorganisms of importance in wastewater treatment are the eukaryotic protists, the procaryotic protists and the virus.

4.2 BACTERIA

Bacteria provide the largest component of the microbial community in all biological wastewater treatment processes, and numbers in excess of 10^6 bacteria/ml of wastewater are frequently encountered (Table 4.1). Because of their large numbers and small size, specialised techniques have been developed to allow ecologists to study these organisms in their natural habitat, in order to provide information which can be used by engineers in plant design and operation. Investigations into bacterial behaviour are generally performed by monitoring organisms in their natural environment while they are growing under changing environmental conditions, in competition with other organisms for a limited food supply. Alternatively, bacteria may be isolated from this environment and studied under carefully controlled laboratory conditions. This latter approach generally proves more simple, and has provided us with much of our information on the growth and nutrition of bacteria.

Because of the small size of bacteria (with the majority having a diameter in the range 0.2–1.5 μm), their identification is impossible by visual means alone. However, by utilising a light microscope of high resolving power, it is possible to note their different shapes and sizes as well as their possession of fine appendages. These features make up the bacterial morphology, and recognition of differences between them forms an important first step in bacterial identification.

Bacterial morphology

Bacteria are limited to four basic shapes a sphere, a straight rod, a curved rod and a spiral. A spherical bacterium is known as a coccus and ranges in diameter from 0.2–4 μm. The cocci may also be grouped according to their spatial arrangement, which is determined by their mode of division. Cocci divide by fission whereby the bacterium

Table 4.1 Numbers of total and viable bacteria at different stages of sewage treatment. Values are geometric means.

Sources (and number) of samples	Bacterial counts				Viability (%)	Total Suspended Solids (mg/l)
	In samples (no./ml)		In suspended solids (no./g)			
	Total	Viable	Total	Viable		
Settled sewage (46)	5.6×10^8	6.3×10^6	3.0×10^{12}	3.4×10^{10}	1.1	190
Activated sludge mixed liquor, conventional rate (18)	5.9×10^9	4.9×10^7	1.3×10^{12}	1.1×10^{10}	0.83	4 600
Activated sludge mixed liquor, high rate (24)	1.4×10^{10}	2.4×10^8	3.0×10^{12}	5.0×10^{10}	1.7	4 800
Filter slimes (18)	6.2×10^{10}	1.5×10^9	1.3×10^{12}	3.2×10^{10}	2.5	54 000
Secondary effluents (16)	5.4×10^7	1.1×10^6	1.9×10^{12}	4.1×10^{10}	2.1	28
Effluents, high-rate activated-sludge plants (24)	4.8×10^7	1.4×10^6	3.3×10^{12}	1.0×10^{11}	3.0	14
Tertiary effluents (11)	2.9×10^7	6.6×10^4	3.0×10^{12}	6.8×10^9	0.23	9.7

(a)   (b)   (c)

(d)   (e)   (f)

Figure 4.2 Common bacterial shapes: (a) typical cocci, (b) diplococci, (c) streptococci, (d) rods, (e) vibrio, (f) spirilla.

increases in size, constricts in the middle and finally breaks in two. An organism which does not split after division, but remains as a pair, is known as a *Diplococcus*, whereas if it continues to split along the same plane resulting in a long chain it is termed a *Streptococcus*. The faecal indicator bacterium *Streptococcus faecalis* is one such organism. Finally, where division takes place in three dimensions at the same time a characteristic cube of eight cocci results, this is known as *Sarcina* (Figure 4.2).

The straight rod represents the commonest bacterial morphology both in natural environments and wastewater treatment plants. It covers many of the frequently encountered genera of bacteria such as *Pseudomonas, Zoogloea, Escherichia* and *Salmonella* (Figure 4.2).

Curved rods possess a single curve which is almost C-shaped, they are known as Vibrio (Figure 4.2). Included in this group of bacteria is the causative agent of cholera, *Vibrio cholera*, and an organism capable of reducing sulphate to sulphide, known as *Desulfovibrio*.

The final bacterial form is a screw or spiral shape known as *Spirillum* (Figure 4.2). These are observed almost exclusively in water samples and routinely observed in samples taken from anaerobic digesters. These curved rods may possess less than one complete turn, to many turns, and the length of the helix can vary from 0.5–60 μm. They are motile bacteria which swim in straight lines with a characteristic corkscrew-like motion. A common causative agent of diarrhoea, *Campylobacter jejuni*, has small tightly-coiled spirals.

The gram stain

In order to render bacteria more visible under a light microscope, a variety of staining techniques are employed to colour either the entire bacterium, or selected external and internal structures. One such stain of fundamental importance to microbiologists, was developed by Christian Gram in 1884. Using his procedure, a drop of a water sample containing bacteria is placed on a slide in a drop, and the water evaporated by gentle heating. This process fixes the bacteria to the slide, and they are then washed with a basic dye such as crystal violet, followed by an iodine wash to fix the stain. This dye is taken up in similar amounts by all bacteria and thus they would appear blue under a microscope. Not all bacteria are capable of retaining this stain however, and after washing with a neutral solvent, such as ethanol or acetone, these bacteria would be decolourised. If a counter stain of a different colour, such as safranine or carbol-fuchsin, is now applied, the decolourised bacteria would take up this stain and appear red. After staining a mixed population of bacteria, and examining under a microscope, those bacteria which have retained the stain would appear blue and are termed gram positive whereas those which were decolourised and counterstained would appear red and are known as gram negative. It is usual to discuss bacterial populations in such terms as gram negative rods or gram positive cocci.

In addition to bacterial shape, certain bacteria also possess specialised appendages which may be visualised using appropriate staining techniques. Confirmation of the presence of these appendages provides valuable information to aid in their identification.

Flagellae and pili

A flagellum is a whip-like structure which provides one of the means by which bacteria achieve motility. The presence or absence of flagella, their number, and their spatial distribution around the bacterium, are all species characteristics. The majority of flagellated bacteria are rod-shaped and three distinct arrangements of flagella are recognised: a single flagellum situated at one or both polar ends (monotrichate); a tuft of flagella at one or both ends (lopotrichate); and large numbers of flagella distributed around the surface of the bacterium (peritrichate). Some examples of flagella are illustrated in Figure 4.3. The number of flagella on an individual bacterium range

Figure 4.3 Arrangement of bacterial flagella: (a) monotrichate, (b) lopotrichate, (c) peritrichate.

(a)

(b)

(c)

from one up to a hundred, and typical swimming speeds of 2 μm/s for peritrichous bacteria and 200 μm/s for monotrichates have been measured.

In addition to flagella, many gram-negative bacteria appear to have a fine covering of short hair when viewed under an electron microscope. These hairs, which are less than 1 μm long and 0.01 μm in diameter, are known as pili or fimbriae. Their exact function appears to be variable as they are implicated in the transfer of DNA between two bacteria. In addition pili confer on bacteria the ability to adhere both to each other as well as to other organisms and this may play some role in the initiation of sludge flocculation.

Extracellular material

When stained with a preparation of Indian ink and examined under a microscope, some bacteria appear to be surrounded by a refractile coating. If this coating is extensive with a clearly defined boundary attached to the cell, it is known as a capsule (Figure 4.4), whereas if it has no clear boundary and appears discrete from the cell, it is referred to as a slime layer (Figure 4.4). Both capsules and slime layers are predominantly polysaccharide in composition. The presence of organisms with these properties appear to be important in the activated sludge process, as this polysaccharide material has been implicated in the mechanism of flocculation of activated sludge. In addition the organism *Zoogloea ramigera*, which produces copious quantities of extracellular polysaccharide, is thought to be one of the first organisms to attach to media in trickling filters. The polysaccharide it produces then acts as a matrix to enmesh other microorganisms. The organisms which produce extracellular material are able to grow equally well whether or not they are producing it, and its role is unclear. It is thought to confer certain survival advantages such as resistance to desiccation and resistance to attack by bacteriophage. It has also been suggested that during periods of nutrient depletion, bacteria might be able to degrade their own slime layer and use it as a source of carbon and energy. Such an ability would prove a useful advantage in wastewater treatment systems where conditions of

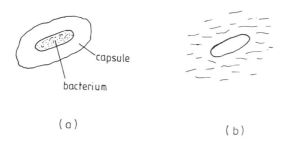

(a)

(b)

Figure 4.4 Bacterial extracellular polysaccharide: (a) in the form of a capsule; (b) As a discrete slime layer.

nutrient depletion are frequently encountered (for instance in the secondary sedimentation tank of an activated sludge plant, or in the maturation ponds of a waste stabilisation pond).

Intracellular structures

The most distinct and characteristic features of bacteria are their shape and extracellular appendages; however, with appropriate staining, many bacteria are seen to contain intracellular structures. In certain cases possession of these structures reflects on the nutritional status of the organism and thus provides information on the influent composition or plant operating conditions. The most common intracellular structure in bacteria of wastewater origin is composed of reserve material. These compounds have an important physiological function, as they are capable of sustaining bacterial growth when there are no extracellular sources of carbon or energy available. Typical reserve materials include: glycogen, polyphosphate and poly-2-hydroxybutyrate (phb). This latter may be discerned either by staining with a lipophilic dye such as Sudan black (hence it is often termed a Sudanophilic granule), or by phase contrast microscopy when it appears refractile (Figure 4.5). Many bacteria found in activated sludge plants possess phb granules, including *Zoogloea ramigera*.

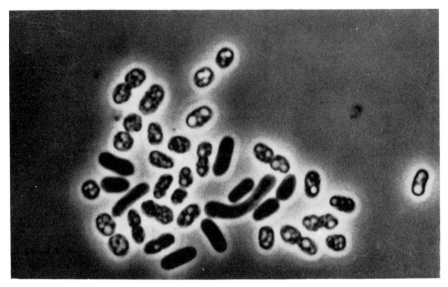

Figure 4.5 Granules of the intracellular storage polymer phb. (Photograph courtesy of E.A. Dawes, University of Hull.)

Another important granular inclusion, is elemental sulphur which results from the bacterial oxidation of sulphide and thiosulphate. Wastewater samples suspected of containing sulphur-storing bacteria are shaken and incubated with sodium sulphide or thiosulphate for 5 min before viewing by phase contrast microscopy. Under these conditions the sulphur appears as clear, circular granules. Sulphur granules are easily discriminated from phb granules since the former occur predominantly in filamentous bacteria (in particular *Beggiatoa* and *Thiothrix*). These are recognised by the length of their individual filaments, which may be over 5000 μm. The presence of large numbers of filamentous bacteria in an activated sludge reactor is an indication of septicity in the sewerage network. This results in the reduction of sulphate in the raw wastewater to sulphide by sulphate-reducing bacteria in the sewers. The sulphide is then oxidised to sulphur under the aerobic conditions of the aeration tank. *Beggiatoa* are a normal component of the flora of a facultative waste stabilisation pond, where a complete sulphur cycle is in operation.

Spores

As a result of changes in environmental conditions, such as nutrient depletion (in particular carbon, nitrogen or phosphorus), or desiccation, many bacteria are able to initiate physiological and morphological changes which result in the formation of dormant structures. One such structure is known as a spore and it is adapted for prolonged survival under adverse conditions for instance heating, desiccation, freezing, toxic chemicals and radiation. Important spore-forming bacteria include *Bacillus anthracis*, which causes the disease anthrax in cattle and sheep, and *Clostridium perfringens* which is used as an indicator of remote faecal pollution. Spores are formed either at the centre or at one end of the cell and are highly refractile. They generally stain poorly and require a powerful stain such as warm fuchsin in phenol. The mother cell in which the original spore was formed (known as a sporangium), disappears once the spore has been produced, and stained preparations of spore-forming bacteria often reveal the various stages of the sporulation process. A dormant structure, similar to a spore, is known as a cyst. It differs in that the bacteria surrounds itself with a tough, protective outer layer and thus no sporangium is involved in its formation.

With a basic knowledge of bacterial morphology, it is possible to attempt an elementary classification scheme of the bacteria. One such scheme, which is widely accepted by microbiologists, divides the bacteria into nineteen divisions of similar characteristics, and those divisions important in wastewater treatment, together with some of their distinguishing properties, are outlined in Table 4.2.

Shape	Oxygen requirements	Spore formation	Flagella	Genera
Gram-positive bacteria				
Cocci	Aerobic	—	—	*Sarcina*
Cocci	Facultative	—	—	*Staphylococcus*
Cocci (long chains)	Facultative	—	—	*Streptococcus*
Rods	Aerobic	+	Peritrichous	*Bacillus*
Rods	Anaerobic	+	Peritrichous	*Clostridium*
Gram-negative bacteria				
Rods	Aerobic	—	Peritrichous	*Nitrosomonas*
				Nitrobacter
Rods	Aerobic	—	Polar	*Pseudomonas*
Rods	Facultative	—	Peritrichous	*Escherichia*
				Salmonella
				Shigella
				Aerobacter
Curved rod	Aerobic	—	Polar	*Vibrio*
Curved	Anaerobic	—	Polar	*Desulfovibrio*

Table 4.2 Morphological properties of bacteria important in wastewater treatment.

+ = Spore formation occurs
− = no spore formation

Algae

Algae are photosynthetic eukaryotes and form a large and heterogeneous group which inhabit fresh, salt and soil water. Only two wastewater treatment processes provide them with a suitable environment namely trickling filters and waste stabilisation ponds, and it is only in the latter that they play a beneficial role in treatment. The criteria which differentiate algae from other Protista is their ability to produce energy from photosynthesis using chlorophyll as the light-gathering apparatus. In addition their nutrition results in the accumulation of organic material, whereas metabolism by the other Protista results in the breakdown of organic material. Algae are classified into a number of divisions primarily on the basis of pigmentation, and this is determined by the nature of the chlorophyll they contain. Other important features used in their differentiation include cellular organisation, cell-wall chemistry, storage products and the presence or absence of flagellation. Lists of algae which have been observed on trickling filters contain over 50 species, and upwards of 150 have been found in waste stabilisation ponds. Examples of typical genera found in these processess are given in Table 4.3. It will be noticed that only four divisions are of importance in wastewater treatment and these are the: Cyanophyta, Euglenophyta, Chlorophyta and the Chrysophyta. The significant characteristics of each of these divisions, together with their distinguishing features and importance, are discussed below.

Division	Genus	Facultative pond	Maturation pond	Trickling filter
Euglenophyta	*Euglena*	+	+	+
	Phacus	+	+	−
Chlorophyta	*Chlamydomonas*	+	+	+
	Chlorogonium	+	+	−
	Pyrobotrys	+	+	−
	Ulothrix	−	−	+
	Stigeoclonium	−	−	+
	Scenedesmus	−	+	−
	Volvox	+	+	+
	Oocystis	−	+	+
	Chlorella	+	+	+
Chrysophyta	*Navicula*	−	+	+
	Bodo	−	+	+
	Oicomonas	−	+	−
Cyanophyta	*Oscillatoria*	+	+	+
	Spirulina	−	+	−
	Anabaena	−	+	−
	Ulothrix	−	−	+

Table 4.3 Algal genera found in wastewater treatment processes.

+ = present
− = absent

Chlorophyta (green algae)

These comprise the major group of algae and are found in aquatic environments ranging from oligotrophic freshwaters through to eutrophic saline waters. They exhibit a wide variety of cellular organisation, although only unicellular and filamentous forms are encountered in wastewaters. Their most distinguishing cellular structure is the chloroplast which is easily visible with a light microscope and is an important criterion in their classification. Many of them are motile by flagella and these flagellae are all of equal length. The end-product of photosynthesis in these organisms is starch and the starch grains are usually associated with a structure called a pyrenoid. Common genera include *Chlamydomas* and *Scenedesmus* (Figure 4.6).

Euglenophyta

These organisms are a group of unicellular flagellates which may be either pigmented or colourless. Pigmented forms possess one long flagellum and a second much shorter one. The colourless forms also have one long flagellum but may have up to three shorter ones. They do not possess a rigid cell wall, but are bounded by a structure known as a plasmalemma and underneath this a protein coat or pellicle. In certain species this arrangement confers pliability which is manifested as a flowing, contracting and expanding movement when the organism is not swimming. Unlike the green algae, Euglenophyta do not store glucose in the form of starch, but as paramylon. Their nutrition is varied but no obligate photoautotrophs are recognised because they have a requirement for vitamins. Organic compounds may be utilised as nitrogen sources and also as an energy source to permit chemotrophic growth in the dark. Typical genera include *Euglena* and *Phacus* (Figure 4.6).

Cyanophyta (blue-green algae)

The prokaryotic blue-green algae are ubiquitous in aquatic environments and are responsible for the large floating masses or 'algal blooms' which accumulate on the surface of pools and lakes. They comprise filamentous, unicellular or colonial species, the latter consisting of individual cells embedded in a polysaccharide matrix. Flagellated stages have not been observed in these organisms, but some filamentous types are capable of a slow, gliding mobility. Although nutrition is mainly photoautrophic, both photohetero-

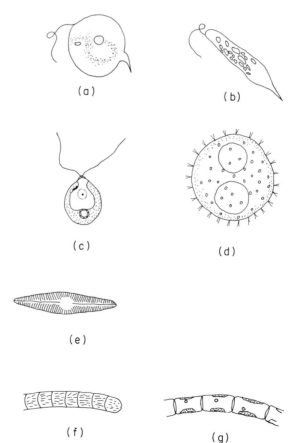

Figure 4.6 Examples of algae associated with trickling filters and waste stabilisation ponds: (a) *Phacus*, (b) *Euglena* (Euglenophyta); (c) *Chlamydomonas*; (d) *Volvox* (Chlorophyta); (e) *Navicula* (Chrysophyta); (f) *Oscillatoria*, (g) *Ulothrix* (Cyanophyta).

trophs and chemoheterotrophs have been observed. Under conditions of reduced oxygen tension, certain of the blue-green algae are capable of 'fixing' molecular nitrogen to ammonia and this frequently exacerbates the problems of nitrogen pollution in eutrophic waters. Many bacteriologists consider the Cyanophyta to be bacteria on the basis of their cellular organisation, and refer to them as Cyanobacteria. The case for them being classified as algae is based on their possession of chlorophyll-*a* and the liberation of free oxygen during photosynthesis, both properties traditionally associated with the algae. They have very short mean generation times which may be as low as 3 hours and reproduction is by cell division. Colonial and filamentous types reproduce by fragmentation whereby the detached segments separate and form new individuals. Individual cells become constricted in the median plane with the synthesis of a new cell wall. *Oscillatoria, Anabaena* and *Microcystis* are typical blue-green algae, all of which are capable of forming algal blooms (Figure 4.6). In addition the latter is associated with the production of a toxin that is poisonous to fish and cattle.

Chrysophyta

This division is of limited importance in wastewater treatment and contains a group of algae which are very diverse with respect to their pigment composition, cell-wall structure and type of flagellated cells. Their preferred habitat is unpolluted freshwater systems and they are more widespread at lower temperatures. The Chrysophyta *Bodo caudatus*, is unusual in that it is found in anerobic waste stabilisation ponds (Figure 4.6).

4.3 PROTOZOA

Protozoa are eukaryotic organisms which demonstrate a wide diversity in form and mode of life. They are generally unicellular, motile and classified on the basis of their morphology, in particular as regards their mode of locomotion. Many protozoa, such as the malarial parasite *Plasmodium*, are parasitic and require the presence of a host organism to complete their life cycle; however, no such protozoa are associated with wastewater treatment processes. Because of their large size, protozoa may be identified easily by inspection under a suitable light microscope and three phyla are to be observed in wastewater treatment processes.

Mastigophora—the flagellated protozoa

These protozoa possess one or more flagella which are used both for locomotion and feeding. They are subdivided into two classes. The

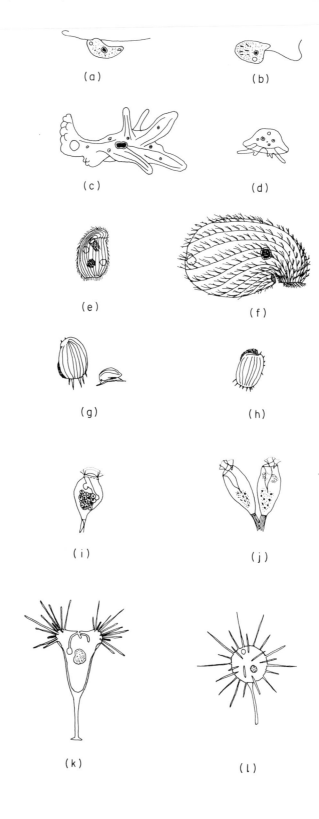

(a)

(b)

(c)

(d)

(e)

(f)

(g)

(h)

(i)

(j)

(k)

(l)

phytomastigophorea usually contain chloroplasts and are capable of autotrophic as well as heterotrophic nutrition. Consequently many organisms classified as phytomastigophorea are also classified as algae, this includes such organisms as *Euglena*, *Volvox*, *Oicomonas* and the dinoflagellates. The remaining class Zoomastigophora, are heterotrophic and have oval-shaped bodies. They are generally the most numerous protozoa in terms of their total numbers in activated sludge plants and trickling filters (Figure 4.7).

Sarcodina—the amoeba

The Sarcodina possess pseudopodial structures which are used for movement and also for feeding, by means of protoplasmic flow. They demonstrate considerable diversity in that some of them lack any skeletal structure (naked amoeba), whereas others have elaborate shells which are known as tests (testate amoeba). The shells are composed of proteinaceous, silicaeous or carbonaceous material, and a wide range of structures have evolved. These shells do not cover the whole body, however, and naked pseudopodia are still used for feeding and locomotion (Figure 4.7).

Ciliophora—the ciliates

The Ciliophora is the largest of the three phyla in terms of the number of species it represents, with over 7000 described in nature. They also provide the greatest species diversity in wastewater treatment plants, although not necessarily the largest number of individuals. They are

---

Figure 4.7 Representative protozoa from the three phyla found in activated sludge and trickling filters:

	Species	Sub-class	Phylum
(a)	*Bodo caudatus*		
(b)	*Oicomonas termo*		Zoomastigophora
(c)	*Amoeba proteus*		
(d)	*Arcella vulgaris*		Sarcodina
(e)	*Chilodenella uncinata*		
(f)	*Colpoda cucullus*	Holotrichia	Ciliophora
(g)	*Aspidisca costata*		
(h)	*Euplotes moebiusi*	Spirotrichia	Ciliophora
(i)	*Opercularia microdiscum*		
(j)	*Vorticella microstoma*	Peritrichia	Ciliophora
(k)	*Podophyra collini*		
(l)	*Acineta tuberosa*	Suctoria	Ciliophora

characterised by the arrangement of cilia in an ordered fashion over the surface of the cell, which serves to effect locomotion. In addition, cilia are also distributed around what is the protozoal equivalent of a mouth, namely the cytosome. Here they provide an aid to feeding by the production of feeding currents (Figure 4.8). Four distinct types of Ciliophora may be identified, based on their locomotion and arrangement of cilia:

1. Holotrichia—free-swimming protozoa which have cilia arranged uniformly over their whole bodies. Typical species are *Trachelophyllum* and *Litonotus*.

2. Spirotrichia—these possess a flattened body with locomotory cilia found mainly on the lower surface. The cilia concerned with feeding are well developed and wind clockwise to the cytosome, they are represented by *Aspidisca, Stentor* and *Euplotes*.

3. Peritrichia—these are immediately recognisable by their inverted, funnel- or bell-shaped bodies which are mounted on a stalk. The other end of this stalk is attached to particulate material such as a sludge floc, and serves to anchor the protozoa. In certain species the stalk is contractile. The wide end of the bell acts as an oral aperture in the Peritrichia and they have cilia arranged around this as an aid in feeding. Typical peritrichs are *Vorticella* and *Opercularia*.

4. Suctoria—the final type of ciliophora is ciliated only in early life, when the cilia enable the young suctoria to disperse from their parents. After this they lose their cilia and develop a stalk and feeding tentacles. The stalk is non-contractile and attaches to particulate material, whereas the feeding tentacle is capable of capturing, and then feeding, on other protozoa. This is achieved by piercing them and sucking in organic material from the cytoplasm to form food vacuoles. Suctoria found in wastewaters include *Acineta* and *Podophyra*.

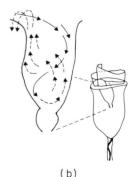

Figure 4.8 Feeding currents induced by (a) flagella, (b) cilia.

(a)

(b)

Protozoal nutrition

Protozoa demonstrate a wide range of feeding strategies of which four types are represented by the protozoa found in wastewater treatment systems. Certain members of the Phytomastigophorea are primary producers and capable of photoautotrophic nutrition, in addition to the more usual chemoheterotrophic nutrition.

Heterotrophy among the flagellated protozoa contributes to the process of BOD removal, and uptake of soluble organic material occurs either by diffusion or active transport. Protozoa which obtain their organic material in such a way are known as saprozoic, and are forced to compete with the more efficient heterotrophic bacteria for the available BOD. Amoebae and ciliated protozoa are also capable of forming a food vacuole around a solid food particle (which include bacteria) by a process known as phagocytosis (Figure 4.9). The organic content of the particle may then be utilised after enzymic digestion within the vacuole, a process which takes from 1 to 24 hours. This is known as holozoic or phagotrophic nutrition, and does not involve direct competition with bacteria, which are incapable of particle ingestion.

The final nutritional mode practised by the protozoa is that of predation. These predators are mainly ciliates, some of which are capable of feeding on algae (and are thus herbivores), as well as other ciliate and flagellate protozoal forms.

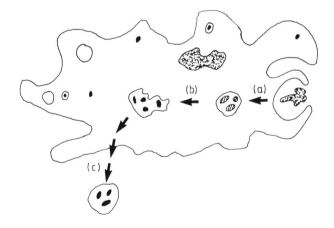

Degradation of a food particle by phagocytosis
(a) food particle engulfed by Pseudopodia and a
    vacuole formed
(b) enzymic digestion occurs within the vacuole and
    digestion products released to the cytoplasm
(c) undigested remains ejected from the body

Figure 4.9 Degradation of a food particle by protozoal phagocytosis.

Within the latter two modes of nutrition, the protozoa display a certain degree of selective feeding. Larger forms of amoebae are carnivorous, eating mainly ciliates and flagellates, whereas the smaller amoebae feed primarily on bacteria. The predatory suctorians are found to feed almost exclusively on holotrichous and spirotrichous ciliates, with hypotrichs, flagellates and amoebae rarely being captured. Peritrichous ciliates are primarily bacterial feeders, but have a limited number of bacterial species upon which they can feed. Certain bacterial species are capable of supporting growth for long periods, whereas others induce starvation after a short time. In addition many bacteria, in particular the pigmented types, prove toxic to those ciliates which ingest them.

4.4 THE ROLE OF MICROORGANISMS IN WASTEWATER TREATMENT PROCESSES

The activated sludge process

The contents of an activated sludge reactor comprise an aerated mass of flocs, surrounded by the influent wastewater or mixed liquor. These activated sludge flocs are made up of aggregates of microorganisms, inorganic and organic colloidal material and larger particulate material, all held together in a compact organic matrix. A large number of protozoa are attached to the floc by means of stalks, and in addition free-swimming ciliates and flagellates are found both in the mixed liquor and within the floc matrix. Although fungi are occasionally observed, they appear to play little part in treatment. Similarly, although bacteriophage are to be found in activated sludge (with counts of coliphage as high as $2 \times 10^4 \, ml^{-1}$), it is likely that their only role is in the removal of bacteria, which will inevitably include a

Genus	Function
Pseudomonas	Removal of carbohydrate, slime production, denitrification
Zoogloea	Slime production, floc formation
Bacillus	Protein degradation
Athrobacter	Carbohydrate degradation
Microthrix	Fat degradation, filamentous growth
Nocardia	Filamentous growth, foaming and scum formation
Acinetobacter	Phosphorus removal
Nitrosomonas	Nitrification
Nitrobacter	Nitrification
Achromobacter	Denitrification

Table 4.4 Reactions carried out by the principal genera of bacteria found in activated sludge.

number of pathogenic species. The microbial population in an activated sludge reactor is highly specialised with a low species diversity, of which the dominant bacteria are gram-negative rods. These are all heterotrophic organisms with the exception of the autotrophic nitrifiers, and the principal genera, along with their postulated role are outlined in Table 4.4.

The main function of conventionally operated activated sludge plants is in the removal of BOD, and this is performed almost exclusively by the heterotrophic bacteria. However, many protozoa are capable of a saprobic mode of nutrition, and activated sludge communities which have reduced protozoal populations, show reduced BOD removal (Table 4.5). The heterotrophic bacteria and saprobic protozoa form the lowest trophic level and act as a food source for holozoic protozoa such as *Euplotes* and *Epistylis*, as well as the rotifers. This would appear to be the most important role for protozoa, in which they act as polishing agents by grazing on free-swimming bacteria. This results in a lower effluent BOD, suspended solids and may contribute significantly to the removal of pathogens (Table 4.5).

Although microorganisms are the agents for BOD removal, the sludge floc has an essential role to play in this process. Up to 40% of the incoming BOD, both soluble and insoluble, is rapidly absorbed to the floc matrix by ionic interactions. Here it is accesible to hydrolysis by extracellular enzymes, before being absorbed and metabolised by the floc bacteria. Since the diameter of a floc varies from 50 to 500 μm, there will be a steep concentration gradient of BOD and oxygen from the outside of the floc, where it will be greatest, to the centre where there may be no residual BOD or oxygen left at all (Figure 4.10). Consequently, towards the centre of a floc, the bacteria will be deprived of a source of nutrient, and thus floc viabilities as low as 1–10% are frequently recorded (Table 4.1).

In addition to its role in promoting BOD removal, a far more important role for the sludge flocs is in promoting good settlement in the secondary sedimentation tanks. Under the quiescent conditions

Effluent analysis	Without ciliates	With ciliates
BOD$_5$ (mg/l)	53–70	7–24
COD (mg/l)	198–250	124–142
Permanganate value (mg/l)	83–106	52–70
Organic nitrogen (mg/l)	14–21	7–10
Suspended solids (mg/l)	86–118	26–34
Optical density at 620 nm	0.95–1.42	0.23–0.24
Viable bacterial count (millions/ml)	106–160	1–9

Table 4.5 The effects of ciliated protozoa on the effluent quality of bench-scale activated sludge plants. (From Curds *et al.*, 1968. Reproduced by permission of the Institution of Water and Environmental Management.)

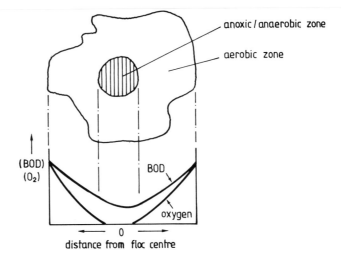

Figure 4.10 The oxygen concentration gradient across a typical activated sludge floc.

of the settlement tank, activated sludge forms large, compact flocs which settle rapidly to yield a sludge with a high solids concentration. This is important as it allows wasting of sludge with a reduced water content, thus reducing sludge-handling charges. In addition, it results in an increased zone settling velocity in the settlement tank, and permits the recycle of a sludge with a higher solids concentration. Why microorganisms undergo flocculation, and the exact mechanisms of this phenomena, are largely unknown. Many models of varying complexity have been proposed, of which the most attractive is the filament backbone model; as well as being a tangible hypothesis, it also provides a working model to allow wastewater engineers to derive plant operation and control strategies. It envisages that, in the reactor, the filamentous bacteria form a matrix or backbone, to which so-called floc-forming bacteria may then attach. This attachment is thought to be brought about by exopolysaccharides, present in the form of a capsule or a discrete slime layer. The archetypal slime-producing bacteria is usually referred to as *Zoogloea ramigera*, although the evidence for this is scant and slime producers are likely to come from a number of genera, principally the Pseudomonads. Continued exopolymer production results in the entrapment of other microorganisms and colloidal particles, and as a consequence the floc diameter increases. Finally the protozoa attach and colonise the floc, and there is some evidence that they too excrete a sticky mucus which may help in strengthening the floc (Figure 4.11). The implications of this model are that if the filamentous and floc-forming bacteria are present in a balanced ratio, then a good floc results. However, if floc-formers predominate, there is insufficient floc rigidity and small, weak flocs are produced which settle poorly. This condition is known as a pin-point floc. Conversely, if filamentous bacteria predominate, then the protruding filaments prevent the close approach of other flocs

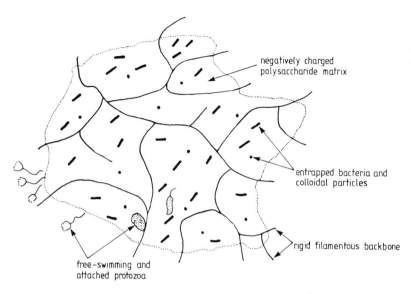

negatively charged
polysaccharide matrix

entrapped bacteria and
colloidal particles

rigid filamentous backbone

free-swimming and
attached protozoa

Figure 4.11 Microbial composition of an activated sludge floc.

during settlement. Thus, upon settlement, the flocs occupy an excessive volume, a condition known as bulking. The phenomena of sludge bulking is discussed in more detail in Chapter 9.

The operation of an activated sludge plant involves the recycling of settled activated sludge. Consequently, microorganisms which are capable of rapid settlement in the sedimentation tank will be recirculated back to the aerator, allowing further division. The organisms which do not settle will therefore either be devoured by grazing protozoa, or discharged with the effluent into a watercourse. This means that there is an inherent selective pressure for flocculating bacteria; free-swimming bacteria must demonstrate extremely high growth rates in order to remain an established population. The intense competition prevailing in the activated sludge process is illustrated by the fact that the majority of the floc flora do not arise from the mammalian enteric tract, despite the continuous high input of bacteria from this source.

Trickling filters

Trickling filter ecology is concerned largely with the composition of the slime layer or biofilm, which adheres to the surface of the medium. It is the species composition of this layer which determines the treatment efficiency. The composition of the biofilm is characterised by a vertical stratification of microbial species, resulting from changes in sewage composition as it passes down the filter. Settled sewage, with a high organic strength and nutrient concentration, is deposited at the top of the filter as a falling film, and this results in a high oxygen concentration due to absorption from the atmosphere. In addition the upper layer of filter media is exposed to light and thus providing conditions which are ideal for algal growth. In the summer months algae may form an extensive green covering on the surface of the filter. Frequently this covering may comprise large filamentous algae such as *Phormidium* or *Stigeoclonium*, and under these conditions the distribution of flow may be impaired and the ventilation decreased.

The process of film formation and colonisation is thought to be initiated by slime-producing and capsulated bacteria which adhere to media surfaces conditioned with an organic film. Attachment probably occurs initially by chemical bonding and van der Waals' forces. It is a very rapid process and *Z. ramigera* is frequently observed as the initial coloniser. Colonisation by other heterotrophic bacteria is also rapid, with *Pseudomonas, Flavobacterium* and *Alcaligenes* among the first to appear. After 5 days biofilm will comprise a diverse assemblage of bacteria, of which filamentous types frequently predominate. Over a period of weeks, the slower growing fungi such as *Fusarium, Geotrichum* and *Sporotrichum* will appear and contribute to BOD removal. In addition algae such as *Chlorella, Euglena, Oscillatoria, Ulothrix* and *Stigeoclonium* will be abundant. Their primary role is likely to be in the removal of nutrients (nitrogen and phosphorus), although in the absence of light it is possible that they may behave heterotrophically and play a small part in BOD removal. Thus a mature biofilm comprises an outer layer which is largely fungi, a middle layer of fungi and algae and an inner layer of bacteria, fungi and algae, and in contrast to the activated sludge process, fungi and algae are the major component of the biomass.

During its passage down the filter, organic material in the wistewater together with oxygen and nutrients, will diffuse into the biofilm and be oxidised by the heterotrophic microorganisms. The

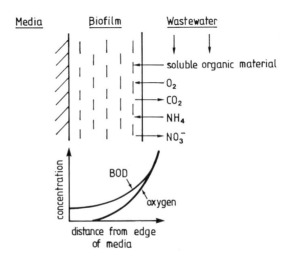

Figure 4.12 Substrate concentration gradients across the slime layer attached to an inert support media in a trickling filter.

amount of organic material and oxygen available for microbial growth will thus depend on the film thickness and the organic load applied to the filter. A point will be reached where the thickness of the film prevents sufficient nutrients reaching the layer of microorganisms which is attached to the medium. Consequently this layer will undergo starvation and ultimately death, causing the entire biofilm to detach from its support (Figure 4.12). This process is known as 'sloughing-off' and the detached film, referred to as humus sludge, is collected in a humus tank.

In addition to horizontal gradients of nutrient depletion across a biofilm, there will also be a steep vertical gradient as BOD is progressively removed. In addition there will be an increase in wastewater alkalinity resulting from dissolved CO_2 and a high oxygen concentration as the oxygen introduced by the draught tubes is not removed by oxidation. Consequently, providing that the filter is not subjected to a high loading, the lower reaches often harbour a large population of autotrophic nitrifying bacteria.

The steep vertical gradient of BOD also influences the distribution of protozoa within a filter. Protozoa are plentiful in filters and over 200 species have been identified. Numerically the ciliates are the most abundant type with from 500 to 10 000 ml^{-1} mixed liquor routinely observed. This compares with 100 to 4600 ml^{-1} for the amoebae and 200–13 000 ml^{-1} for the flagellates. Such information, however, provides little information as to the relative proportions of total biomass, since they are considerably larger than the other component microorganisms. The surface of trickling filters shows a restricted protozoal fauna, comprising mainly the holozoic amoebae and flagellates, whereas a greater variety are found in the lower regions and these are predominantly carnivorous.

Waste stabilisation ponds

The microbial ecology of a waste stabilisation pond is extremely complicated as it closely approximates that of a natural ecosystem. Consequently it has a wide species diversity and incorporates a number of complete nutrient cycles (namely carbon, nitrogen and sulphur). Ponds are unique among biological treatment processes in that, with the exception of anaerobic ponds, primary producers predominate; thus there is an increase in energy through the system. As the functions of anaerobic, facultative and maturation ponds are quite different, then the microbial communities which inhabit these ponds, will also be quite different.

1. Anaerobic ponds

Wastewater entering anaerobic ponds is initially metabolised by a group of facultatively anaerobic heterotrophs. These organisms enzymically hydrolyse polymeric material, such as protein, fats and carbohydrate, into their constituent monomers. The high loading rates applied to anaerobic ponds ensures that oxygen is utilised more rapidly than it is replaced by atmospheric diffusion. Consequently, these monomers may be metabolised in the absence of oxygen resulting in the production of volatile fatty acids and in particular acetate. For this reason the bacteria which catalyse this fermentation, are generally referred to as the acid formers, and typical genera include *Clostridium, Propionibacterium* and *Bacteroides*. The methanogenic bacteria (Methanobacteriaceae) are strictly anaerobic and killed rapidly by relatively short exposure times to oxygen. They obtain their energy for growth via the formation of methane from the reduction of CO_2 and the oxidation of only a limited range of compounds, namely hydrogen, formate, acetate and methanol. Successful operation of an anaerobic pond is dependent upon the establishment of an active population of methanogenic bacteria. These bacteria are extremely fastidious and susceptible to changes in temperature and pH. If the loading rate is too high, volatile acid production exceeds the rate at which they are degraded by the mèhanogens, consequently the pH falls and the methanogen population is killed. In the absence of the methanogens, digestion and stabilisation of a waste is prevented and volatile fatty acids will accumulate. These compounds, in particular butyric acid, are responsible for noxious odours, and the waste is putrefied rather than stabilised.

Another important group of organisms present in anaerobic ponds are the sulphate-reducing bacteria. These are an extensive and varied group of microorganisms which can respire anaerobically, reducing sulphate to sulphide (e.g. *Desulfotomaculum*) or can reduce elemental sulphur to sulphide (*Desulfuromonas*). The range of substrates which they are capable of oxidising is also quite varied, whereas some of them exploit acetate and hydrogen, others grow on higher chain fatty acids and excrete acetate as an end-product. Sulphate reducers are generally beneficial in that they are responsible for the removal of wastes which have a high sulphate concentration. However, production of hydrogen sulphide gas may result in odours if the sulphate content of the influent is excessive.

2. Facultative ponds

The mechanism for the bacterial removal of BOD and ammonia in facultative ponds is identical to that in the activated sludge process and trickling filters. The essential feature of facultative ponds, however, is that the oxygen necessary to satisfy the BOD demand of the waste is produced by algae as a result of photosynthesis, and utilised by the heterotrophic and autotrophic bacteria for the oxidation of carbon and nitrogen. The dominant bacterial genera include *Pseudomonas*, *Flavobacterium* and *Achromobacter*. In return the algae exploit the CO_2 and NH_3 produced from these reactions as a source of cell carbon and nitrogen (Figure 4.13). Although fungi are routinely observed in ponds in small numbers, little is known of their role and ecology. It is also suspected that many of the fungal species isolated derive mainly from the surrounding soil and plants. The ecology of protozoa in ponds reveals similar gaps in our knowledge; although these organisms are present, little is known concerning the types of species or their numbers. It is possible that protozoa may play a significant role in the removal of both viral and bacterial pathogens, although there is no evidence for this at the present time.

As with natural ecosystems, ponds demonstrate a decrease in species diversity with increasing saprobity or organic strength. Thus it is not uncommon to find facultative ponds which contain only a few species of algae with one species dominant. The dominant species is generally a motile algae which is better able to exploit the available surface light by forming dense bands; these are capable of moving up and down the water column to reside at the depth of optimal light intensity. Consequently, flagellated organisms such as *Chlamydomonas* or *Chlorella* tend to predominate. Algal bands with counts of *Chlorella* as high as $5000 \, ml^{-1}$ have been recorded. Vertical speciation in facultative ponds is marked, as there will be a decreasing gradient of light intensity, pH, dissolved oxygen concentration and organic strength. Consequently the upper and middle layers will witness the maximum activity in terms of BOD and nutrient removal. The lower layer of a facultative pond will have a similar ecosystem to that of an anaerobic pond. In addition, however, it is important in the slow anaerobic degradation of large recalcitrant molecules and dead algal cells, with consequent nutrient recycling. This process results in the production of hydrogen sulphide from the reduction of sulphur (which is present in proteins) and sulphate (which escaped reduction in the anaerobic pond), by the sulphate-reducing bacteria. This sulphide is then oxidised to elemental sulphur or thiosulphate by aerobic sulphide oxidisers such as *Beggiatoa* which are present in the aerobic layer.

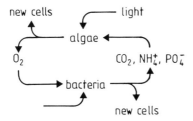

Figure 4.13 Symbiosis of algae and bacteria in waste stabilisation ponds.

3. Maturation ponds

Maturation ponds have a very low organic content and consequently a higher species diversity, resulting in the presence of many more algal types. In addition the increased clarity of the water results in a greater light penetration, which allows non-motile algae such as *Scenedesmus* and *Chlorella* to compete successfully with the flagellates. The main algal genera found in these ponds are listed in Table 4.4. The reduced organic strength of maturation ponds causes a reduction in the total heterotrophic population, although the species diversity increases. Little information is available on the composition of the microbial population, but it might be expected that an increased number of nitrifying bacteria will be present. Despite the high prevalent dissolved oxygen, bacteria play little or no part in the role of maturation ponds, which are primarily for pathogen destruction, nutrient removal and suspended solids removal. Nutrition removal is achieved mainly by the algae which take up large amounts of nitrogen and phosphorus. In addition they aid in pathogen removal by providing an elevated pH as a result of CO_2 stripping. The major mechanisms for the removal of pathogens are as yet unknown. Nutrient starvation which results from the inability of gut microflora to compete successfully for organic substrates present at low concentrations, is likely to prove important. This process is aided by the long retention times in maturation ponds, and will be exacerbated at elevated temperatures due to a higher energy demand necessary to satisfy an increased endogenous respiration rate. Other factors such as ultraviolet (UV) light, oxygen toxicity and the production of algal toxins are also likely to play minor roles.

5 Microbial Energy Generation

5.1 MECHANISMS OF ENERGY GENERATION

The microorganisms which make up the community of a biological reactor are important solely for the chemical and physical changes which they cause. These changes are a direct result of the nutritional requirements and biochemical activities of the component organisms. The exact composition of the community will be determined by those organisms which are able to demonstrate the most efficient growth on the chemical components available in the wastewater and under the prevalent environmental conditions. Growth efficiency in a given environment is a function of two factors. If an organism is to grow it must compete successfully for the limited supply of nutrients which are present in the wastewater and transport them across the cell wall. It must then degrade these compounds in order to produce energy— the greater its yield of energy from a given compound then the greater will be its ability to outgrow its competitors.

Microbial energy generation

All organisms expend energy, not only for growth and reproduction but also to perform many functions not associated with growth. These 'maintenance' functions include such phenomena as motility, the transport of organic and inorganic material across the cell wall and the synthesis of new cell material. The energy derives from either physical energy—namely light, or from the breakdown of chemicals (known as substrates) to release chemical energy. Physical and chemical energy are converted into biologically utilisable energy by microbial metabolism and stored inside the organism in a chemical form as a compound called adenosine 5'-triphosphate (ATP). It consists of an adenosine molecule which is linked to three inorganic phosphate molecules by phosphoryl bonds (Figure 5.1) and the formation of one of these bonds requires the input of a large amount of energy. The energy which results from hydrolysing this bond may be utilised by an organism to perform useful work, and energy produced from chemical reactions which is used in this way is termed 'free energy' with the symbol G. If the chemical reaction yields energy then the free energy change (ΔG), is given a negative sign since the

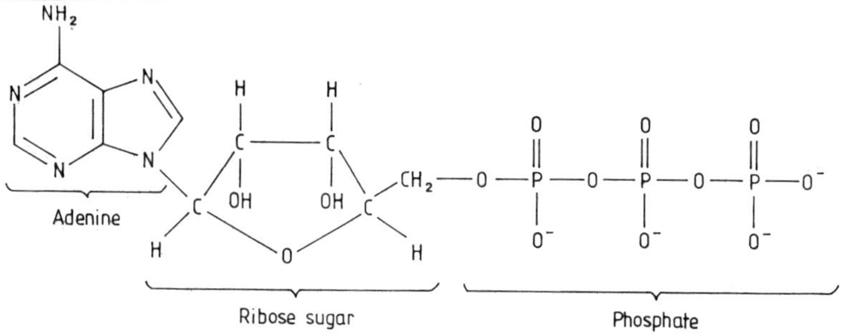

Figure 5.1 The structure of the ATP molecule.

reaction has lost energy to the environment; the converse applies to reactions which require an energy input. The energy resulting from metabolism, which is stored within an organism, is said to have been conserved. Compounds in which this energy is conserved in the form of a phosphoryl bond, are often termed 'high energy' because of the large change in free energy which occurs on hydrolysis. Production of ATP is from the reaction of adenosine 5'-diphosphate and inorganic phosphate with formation of a new phosphoryl bond, thus:

$$ADP^{3-} + P_i + H^+ \longrightarrow ATP^{4-} + H_2O \qquad \Delta G = +30\,\text{kJ/mol}$$
$$(5.1)$$

The symbol ΔG denotes that reaction (5.1) requires an input of 30 kJ of energy to synthesise 1 mol of ATP from ADP and inorganic phosphate $(H_2PO_4^{2-})$, which is generally written in shorthand form as P_i. Provision of this energy comes from a large number of energy-yielding chemical reactions collectively termed catabolism.

Once ATP has been produced, it may be stored inside the cell and used as and when required. The stored energy is released by hydrolysing the phosphoryl bond in a reaction which is the reverse of reaction (5.1):

$$ATP^{4-} + H_2O \longrightarrow ADP^{3-} + P_i + H^+ \qquad \Delta G = -30\,\text{kJ/mol}$$
$$(5.2)$$

This reaction yields 30 kJ of energy for each mole of ATP hydrolysed, and both reactions (5.1) and (5.2) are catalysed by an enzyme called ATPase. The resultant energy can now be used for growth, reproduction and maintenance purposes by a series of reactions termed 'anabolism'.

Redox reactions

The major reactions of catabolism are oxidation reactions and biological oxidations generally involve the removal of hydrogen or electrons. Oxidation reactions are associated with the release of free energy thus the oxidation of a compound AH_2 would proceed as follows:

$$AH_2 \longrightarrow A + 2H^+ + 2e^- \qquad \Delta G = -\text{ve} \qquad (5.3)$$

Similarly, reduction reactions require an input of energy to drive them and the reduction of a compound B would be

$$B + 2H^+ + 2e^- \longrightarrow BH_2 \qquad \Delta G = +\text{ve} \qquad (5.4)$$

Of course, neither an oxidation nor a reduction can proceed in isolation and an oxidation reaction must be coupled to a reduction. When this occurs the reaction is generally denoted as

$$AH_2 \underset{A}{\overset{B}{\diagdown}}\!\!\diagup\!\!_{BH_2} \qquad (5.5)$$

and termed a redox reaction. Reactions of the form AH_2/A and B/BH_2 are known as half-cell reactions or redox couples. Reaction (5.5) helps to envisage the two important features of redox reactions, namely that an oxidation is accompanied by a reduction and that the two are coupled by the transfer of hydrogen or electrons which are often referred to as reducing equivalents. Biological redox reactions are arranged such that they proceed with a net yield of free energy. Inspection of the half-cell reaction (5.3) above, shows that it can undergo either an oxidation by the donation of reducing equivalents $(A \rightarrow AH_2)$ or a reduction by the acceptance of reducing equivalents $(AH_2 \rightarrow A)$. The propensity of a couple to undergo either oxidation or reduction is known as its redox potential, measured in volts. The redox potentials of couples which are of interest in biological systems range from $-0.42\,\text{V}$ to $+0.82\,\text{V}$ (Table 5.1). When two couples are combined the direction of the reaction is decided by the magnitude of the redox potentials of each couple, and a couple of lower redox potential will lose its reducing equivalents to a couple of higher redox potential. Thus from Table 5.1, if the couples $NAD^+/NADH$ and O_2/H_2O are combined, then reducing equivalents will pass from NADH to O_2 and the resulting reactions will be

$$NADH + H^+ \underset{NAD^+}{\overset{O_2}{\diagdown}}\!\!\diagup\!\!_{H_2O} \qquad (5.6)$$

The net energy resulting from the pairing of two redox couples may be calculated from the equation:

$$\Delta G^{0\prime} = -nF\,\Delta E_0' \qquad (5.7)$$

where $\Delta G^{0\prime}$ is the standard free energy change (kJ/mol), n is the number of electrons transferred in the reaction, $\Delta E_0'$ is the change in redox potential (V) and F is the Faraday constant (96.649 kJ/V mol)

Thus for reaction (5.6) above the resulting energy yield would be

$$\Delta G^{0\prime} = -2 \times 96.649 \times [0.816 - (-0.32)] = -219.58\ \text{kJ/mol}$$

Half-reaction	E'_0 (pH 7.0, 30 °c) mV
$^{1/2}O_2 + 2H^+ + 2e \longrightarrow H_2O$	$+816$
$Fe^{3+} + e \longrightarrow Fe^{2+}$	$+771$
Cytochrome a $Fe^{3+} + e \longrightarrow$ Cytochrome a Fe^{2+}	$+290$
Cytochrome b $Fe^{3+} + e \longrightarrow$ Cytochrome b Fe^{2+}	$+80$
Fumarate $+ 2H^+ + 2e \longrightarrow$ Succinate	$+31$
$FAD + 2H^+ + 2e \longrightarrow FADH_2$	-60
Acetaldehyde $+ 2H^+ + 2e \longrightarrow$ Ethanol	-163
Pyruvate $+ 2H^+ + 2e \longrightarrow$ Lactate	-190
$NAD^+ + 2H^+ + 2e \longrightarrow NADH + H^+$	-320
$NADP^+ + 2H^+ + 2e \longrightarrow NADPH + H^+$	-320
$CO_2 + 2H^+ + 2e \longrightarrow$ Formate	-420
$H^+ + e \longrightarrow {}^{1/2}H_2$	-420
Acetate $+ 2H^+ + 2e \longrightarrow$ Acetaldehyde	-600

Table 5.1 Standard electrode potentials of redox half-reactions, of biological importance.

If two half-cell reactions are coupled with direct electron transfer, then the available energy is lost to the organism and dissipated as heat. If however, the half-cell reactions are connected by an intermediate, the energy can be conserved and used to do useful work. Catabolic energy-generating processes in microorganisms achieve this coupling in two ways, either by substrate-level phosphorylation or chemiosmotic energy generation.

1. Substrate-level phosphorylation. This mechanism is particularly important for bacteria which grow in the absence of oxygen and a great many such anaerobes synthesise ATP exclusively by this mechanism. Only six reactions are capable of generating ATP by substrate-level phosphorylation (Table 5.2) and it will be noticed that five of these involve compounds which contain energy-rich phosphoryl bonds.

2. Chemiosmotic energy generation. As mentioned earlier, biological oxidations are performed by the removal of hydrogen or electrons. An acceptor for these reducing equivalents is thus required and the two most common acceptors in biological systems are nicotinamide adenine dinucleotide (NAD^+) and its phosphorylated product $NADP^+$. These two acceptors allow a single oxidation reaction to provide reducing equivalents for many reductive processes. As the reactions of cell synthesis and maintenance (anabolism) are primarily reductive, the NAD^+ and $NADP^+$ facilitate the direct coupling of catabolism and anabolism (Figure 5.2). However, in growing cells catabolic activity is in excess of anabolic activity and an

Table 5.2 Reactions yielding ATP by substrate-level phosphylation in anaerobes.

Reaction			Enzyme	$\Delta G'_{obs}$ (kcal/mol)
1,3-Bisphosphoglycerate + ADP	$\rightleftharpoons$	3-Phosphoglycerate + ATP	Phosphoglycerate kinase	-5.8
Phosphoenolpyruvate + ADP	$\rightleftharpoons$	Pyruvate + ATP	Pyruvate kinase	-5.7
Acetyl phosphate + ADP	$\rightleftharpoons$	Acetate + ATP	Acetate kinase	-3.1
Butyryl phosphate + ADP	$\rightleftharpoons$	Butyrate + ATP	Butyrate kinase	-3.1
Carbamyl phosphate + ADP	$\rightleftharpoons$	Carbamate + ATP	Carbamate kinase	-1.8
N^{10}-Formyl FH$_4$ + ADP + P	$\rightleftharpoons$	Formate + FH$_4$ + ATP	Formyl FH$_4$ synthetase	$+2.0$

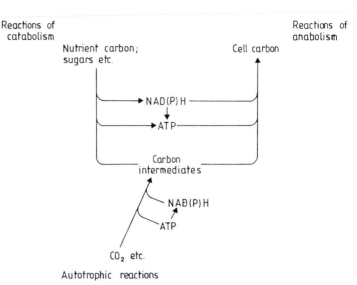

Figure 5.2 The coupling of the energy-yielding reactions of catabolism, and the energy-requiring reactions of anabolism, by the transfer of reducing equivalents via NAD(P)H. (From Lynch and Poole, 1978. Reproduced by permission of Blackwell Scientific Publications Ltd.)

excess of reducing equivalents are produced. As NAD^+ and $NADP^+$ are present only in catalytic amounts, then some other method of reoxidising NAD(P)H back to $NAD(P)^+$ must be available or else catabolism would cease. One of the most widespread mechanisms for this is known as respiration. In this process the reducing equivalents are passed from NAD(P)H through a series of electron carriers to an ultimate electron acceptor. This series of carriers are redox couples of increasing redox potential (Figure 5.3), collectively known as an electron transport chain, and its components are all protein molecules which are attached to the inner cell membrane of the prokaryotic microorganism. In eukaryotes the electron transport chain is found

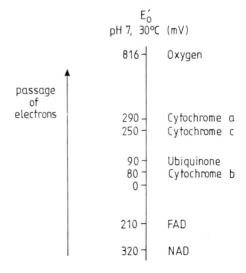

Figure 5.3 The electron carriers of the electron transport chain and their associated redox potentials.

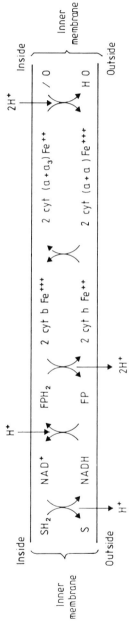

Figure 5.4 A typical bacterial electron transport chain, showing the spatial arrangement of carrier proteins, which produces a separation of electrical charge and pH gradient across the bacterial membrane.

within the mitochondria, an organelle associated with energy generation. A typical bacterial electron transport chain is shown in Figure 5.4, although there are many variation in the exact composition of the chain between prokaryotes and eukaryotes and also between different prokaryotic species. In the presence of oxygen the ultimate redox couple is $\frac{1}{2}O_2 + 2H^+ + 2e^- \rightarrow H_2O$, oxygen is therefore known as the terminal electron acceptor. However, in the absence of oxygen, certain genera of bacteria are capable of using alternative terminal electron acceptors such as nitrate (NO_3^-/N_2) and sulphate (SO_4^{3}/S_2). This is known as anaerobic respiration and is of profound importance in wastewater treatment. The oxidation of 1 mol of NAD(P)H yields sufficient energy for the synthesis of 3 mole of ATP and this coupling occurs at three distinct sites on the electron transport chain (Figure 5.5).

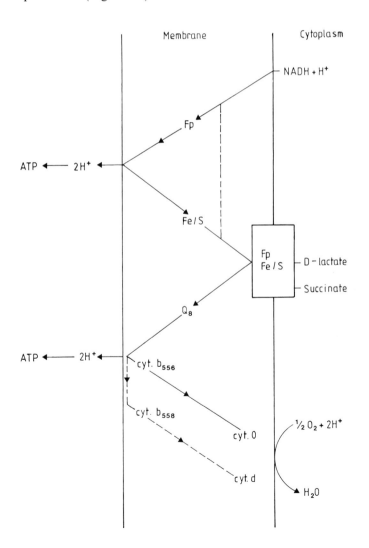

Figure 5.5 Another form of bacterial transport chain, illustrating the sites at which ATP is synthesised. (From Haddock and jones, 1977. Reproduced by permission of the American Society for Microbiology.)

The exact mechanism of coupling was proposed by Peter Mitchell in 1966 and is known as the chemiosmotic hypothesis. This theory proposes that the inner membrane of the eukaryotic mitochondria and the prokaryotic cell, is impermeable to the passage of ions including H^+ and OH^-. As a result of the spatial orientation of the protein carrier molecules in the chain, the transfer of reducing equivalents from NAD(P)H to O_2 results in a separation of both electrical charge and chemical ions. Consequently there is a gradient of pH and electrical potential across the cell membrane with one side alkaline and negative with respect to the other (Figure 5.5). The enzyme ATPase which catalyses reactions (5.1) and (5.2) is also located in the microbial inner membrane and this enzyme is the means whereby the energy stored as an electrochemical gradient, is released; ATPase is capable of coupling the movement of H^+ across the membrane to ATP synthesis, with a resultant neutralisation of the gradient:

$$2H^+_{inside} + ATP \rightleftharpoons ADP + P_i + 2H^+_{outside} \qquad (5.8)$$

The above reaction is fully reversible and thus, when necessary, an electrochemical gradient can be established by the degradation of ATP to ADP (Figure 5.6). The sum of the H^+ gradient (ΔpH) and the electrical or membrane potential ($\Delta\Psi$) is known as the proton motive force and they are all related by the expression:

$$\Delta P = \Delta\Psi - \frac{2.3RT}{F}\Delta pH \qquad (5.9)$$

where R is the gas constant (8.314 J/K mol), T is the absolute temperature, F is the Faraday constant and ΔP is the proton motive force (mV).

Typical values for ΔP are in the range -100 to 300 mV. In addition

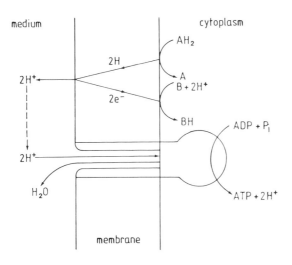

Figure 5.6 Representation of the enzyme ATPase and the mode in which it synthesises ATP by the translocation of protons across the membrane, thus utilising the energy of oxidation of AH_2 by means of the proton motive force.

Figure 5.7 The central role of the proton motive force and its components, in the storage and provision of energy, for the reactions of bacterial metabolism. (From Konings and Veldkamp (1983). *Microbes in their Natural Environments, Society for General Microbioloy Symposium 34,* Cambridge University Press.)

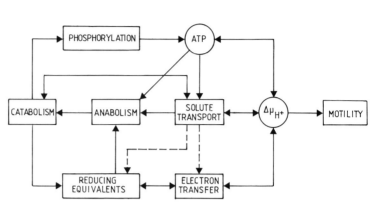

to its role in ATP synthesis, this gradient is also responsible for providing the energy to drive the transport of substrates across the microbial cell wall. The central role played by the electrochemical gradient in bacterial metabolism is illustrated in Figure 5.7.

5.2 PHOTOTROPHIC ENERGY GENERATION

Many organisms are capable of trapping the energy of sunlight and using it to drive the synthesis of ATP. There are two quite distinct types of photosynthesis which are dependent upon the source of reducing power exploited by the organism. The green plants, algae and the Cyanophyta (cyanobacteria) catalyse the oxidation of water to yield oxygen, electrons and hydrogen ions:

$$2H_2O \longrightarrow O_2 + 4e^- + 4H^+ \qquad (5.10)$$

This reaction is coupled to the reduction of CO_2 resulting in the production of carbohydrate. The overall reaction is

$$2H_2O + CO_2 \longrightarrow (CH_2O) + H_2O + O_2 \qquad (5.11)$$

As oxygen is produced in this reaction it is termed 'oxygenic photosynthesis'. By contrast photosynthetic bacteria do not oxidise water but replace this with some other reductant such as S^{2-}, $S_2O_3^{2-}$ or H_2. These reactions do not result in the production of oxygen and are termed 'anoxygenic photosynthesis'. Obligate photosynthetic bacteria are also anaerobic and oxygen is inhibitory to their growth. Photosynthetic prokaryotes fall into three well-defined and distinct groups: the cyanobacteria, the purple bacteria and the green bacteria (Table 5.3).

Table 5.3 Characteristics of photosynthetic bacteria.

Group	Electron donor (substrate for growth)	Growth conditions and other properties	Examples
Green bacteria			
(a) Chlorobiaceae	H_2S $Na_2S_2O_3$ H_2	Light, autotrophic; non-motile	*Chlorobium limicola*, *C. thiosulfatophilum*, *Procochloris aestuarii*
(b) Chloroflexaceae	Organic substrates	Facultative aerobic; filamentous gliding	*Chloroflexus aurantiac*
Purple sulphur bacteria (Chromataceae)	H_2S, $Na_2S_2O_3$ H_2 Organic substrates, e.g. acetate	Autotrophic and heterotrophic in light. Strict anaerobes	*Chromatium* sp. *Thiocapsa roseopersicina*
Purple non-sulphur bacteria (Rhodospirillaceae)	Organic substrates, e.g. succinate, malate H_2	Heterotrophic or autotrophic in light and anaerobic. Will grow aerobically and heterotrophically in the dark	*Rhodospirillum rubrum* *Rhodopseudomonas viridish*

Oxygenic photosynthesis

The location of the apparatus of photosynthesis in eukaryotes is found within chloroplasts (Figure 5.8). This organelle is analogous to the eukaryotic mitochondria as they are both sites of the electron transport chain and energy generation. In the cyanobacteria it is located on the cell membrane. The site of photosynthesis contains many light-harvesting pigments such as chlorophylls, carotenoids and phycobilins which are grouped into photosynthetic units each consisting of 300–400 molecules of pigment. The pigments are capable of absorbing light over a broad range of wavelengths and referred to collectively as antenna pigments. Each photosynthetic unit contains two photosystems (I and II). Photosystem I contains a chlorophyll pigment called P700, which absorbs light maximally at a wavelength of 700 nm. Photosystem II contains a form of chlorophyll which absorbs maximally at 680 nm called P680.

The first stage of photosynthesis is the capture of radiant energy by the antenna pigments which then aid in the migration of this energy to the reaction centres. At the reaction centres, light absorption by P700 and P680 causes them to lose an electron which is transferred to a primary acceptor. There are now two mechanisms available for coupling the transfer of these electrons to ATP synthesis. Non-cyclic photophosphorylation results in the accumulation of NADPH and the synthesis of ATP and it involves both photosystems I and II. Cyclic photophosphorylation results solely in ATP synthesis and involves only photosystem I.

(a) Non-cyclic photophosphorylation

Electron transfer from water to $NADP^+$ requires the photochemical reactions of both photosystems I and II which are linked by a

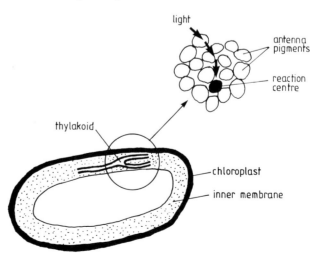

Figure 5.8 Section of a chloroplast containing the thylakoids which are arranged in stacks or grana. The grana are interconnected by membranes known as stroma. Photosystem I particles are found mainly in the lamella and photosystem II particles in the grana.

mechanism known as the 'Z-scheme'. The excitation of P680 by light at photosystem II causes it to emit an electron which is accepted by a carrier known as plastoquinone (PQ) and this becomes reduced. In addition the excited P680 becomes a strong oxidant:

$$P680 \longrightarrow P680^+ + e^- \qquad (5.12)$$

$$PQ + e^- \longrightarrow PQ^- \qquad (5.13)$$

A further two carriers link the transfer of this electron from photosystem II to photosystem I where it is used to rereduce the excited form of P700 (P700$^+$). The passage of four electrons down the chain from P680 to P700$^+$ is associated with the synthesis of four molecules of ATP.

The strong oxidant P680$^+$ now oxidises water at photosystem II by the removal of hydrogen to produce oxygen and a reductant:

$$H_2O \longrightarrow 2H^+ + 0.5O_2 + 2e^- \qquad (5.14)$$

The electrons released in this reaction are used to replace those lost by excitation of P680 and the hydrogen ions are used in the reduction of NADP$^+$ as follows. As a result of excitation of P700 by light, a weak

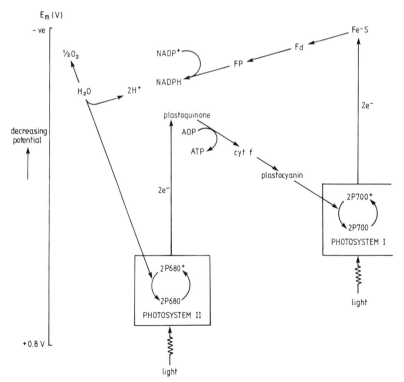

Figure 5.9 The electron transport system and the passage of electrons associated with ATP synthesis by non-cyclic photophosphorylation. The protein carriers are: FP—flavoprotein; Fd—ferrodoxin; Fe-S—iron sulphur protein; Cyt—cytochrome.

oxidant, P700$^+$, is produced and an electron emitted. This electron is passed through a series of three carriers to NADP$^+$ which is then reduced to NADPH using the hydrogens from water. NADPH is now available for CO_2 fixation by the Calvin cycle, and four electrons are required for each molecule of CO_2 fixed.

The essential feature of photosynthetic energy generation is that the excited forms of P680 and P700 can act as reducing agents at much lower redox potentials (1 V more negative) than in the ground state. Thus in the ground state P700 has a redox potential of +0.4 to 0.5 V, whereas after excitation, its redox potential is −0.5 V and it is now capable of reducing NADP$^+$ which has a redox potential of 0.32 V. The energy available therefore, is represented by the difference in redox potential between the excited chlorophyll molecule and the reduced acceptor molecule. The pathway of non-cyclic photophosphorylation is illustrated in Figure 5.9.

(b) Cyclic photophosphorylation

Illuminated chloroplasts are capable of synthesis of ATP even in the absence of added electron donors or acceptors. This reaction involves no evolution of oxygen and NADPH is not produced. It appears therefore that the mechanism involves only photosystem I in which the electron emitted from excited P700 is returned back via a circular chain of carriers. The passage of electrons around this transport chain is associated with ATP synthesis and is illustrated in Figure 5.10.

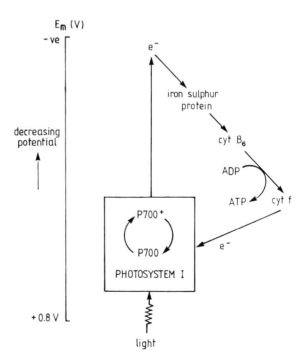

Figure 5.10 The electron transport chain and the passage of electrons associated with ATP synthesis by cyclic photophosphorylation in photosynthetic bacteria.

Anoxygenic photophosphorylation

Anoxygenic photophosphorylation is carried out solely by pro-karyotes of which three families are important in waste stabilisation ponds. These organisms differ from the Cyanophyta and eukaryotic phototrophs in the nature of their antenna pigments which comprise a variety of bacteriochlorophylls and carotenoids. Energy harvested by these pigments is transferred to a reaction centre which contains a bacteriochlorophyll absorbing at 870 (P870) and 890 nm (P890). Only one light reaction (photosystem I) is involved and the electrons lost during the formation of NADH are replaced by oxidation of a narrow range of reduced compounds found in the wastewater.

The Chromatiaceae (purple sulphur bacteria) and the Chloro-biaceae (green sulphur bacteria) are obligate phototrophs and obligate anaerobes which oxidise reduced sulphur compounds such as thiosulphate and hydrogen sulphide to replace electrons lost from the cyclic system. As a result of this the reduced sulphur compound is oxidised to elemental sulphur:

$$2H_2S + CO_2 \longrightarrow (CH_2O) + 2S + 2H_2O \tag{5.15}$$

Rhodospirillaceae (purple non-sulphur bacteria) are primarily photoheterotrophic organisms which also possess the ability to fix CO_2 using hydrogen as reductant. The synthesis of the photosynthe-tic pigments does not occur in the presence of oxygen, but under these conditions they are capable of aerobic chemoheterotrophic growth. Typical organic substrates for the replacement of electrons lost during photophosphorylation include succinate and malate.

5.3 UPTAKE OF SUBSTRATES INTO THE MICROBIAL CELL

Potential substrates for microbial growth are found in both soluble or particulate forms in a wastewater. Although the influent to a biological reactor has been subjected to primary sedimentation, as much as 30% of the BOD is still in an insoluble or colloidal form with particle sizes over the range $1-500\,\mu m$. Such particulate substrates comprise protein, fats and carbohydrates and are directly available to those protozoa capable of phagocytosis (p. 123). As bacteria are surrounded by a rigid cell wall this does not permit the uptake of such large molecules. This source of substrate is not unavailable to them, however, and some species are capable of secreting enzymes extra-cellulary. These have the ability to hydrolyse the larger molecules into smaller soluble molecules which will easily cross the bacterial cell wall. The slime layer in trickling filters and the floc in the activated sludge process play important roles in this mechanism as they permit the entrapment or adsorption of particulate material which facilitates attack by extracellular enzymes.

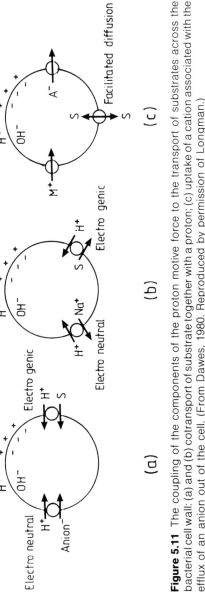

Figure 5.11 The coupling of the components of the proton motive force to the transport of substrates across the bacterial cell wall: (a) and (b) cotransport of substrate together with a proton; (c) uptake of a cation associated with the efflux of an anion out of the cell. (From Dawes, 1980. Reproduced by permission of Longman.)

As a result of its lipid bilayer, the bacterial cell wall presents an extremely hydrophobic barrier to its external environment. Only those compounds which are themselves lipid soluble are able to cross it unaided. If a concentration gradient of such a compound exists across the bacterial cell membrane, with a higher concentration on the outside of the cell, then the substrate may enter the cell by diffusion down the concentration gradient. This process is known as passive transport and does not require the input of energy. It is likely to be unimportant in wastewater treatment processes as the concentration of substrate in the reactor is extremely low by comparison with that inside a bacterial cell.

The majority of substrates rely on the mechanism of active transport for their uptake. Active transport is an energy-requiring process which exploits the electrochemical gradient to drive substrate transport. A microorganism must transport many substrates across its cell wall in order to fulfil its nutritional requirements. These include cations (such as Ca^{2+} and Mg^{2+}) which carry a positive charge, anions (such as SO_4^{2-} and PO_3^{4-}) which carry a negative charge and compounds which carry no charge such as glucose and fructose. The exact mechanism by which the energy conserved as a proton motive force is coupled to substrate uptake is dependent on the charge carried by the substrate. Some examples of this coupling are shown in Figure 5.11.

In order for the substrate to cross the impermeable bacterial cell wall, a group of carrier enzymes (frequently termed permeases) are required. These carriers are located in the cell wall and recognise a specific substrate or group of structurally related substrates. They then bind to their specific substrate and transport it across the membrane. Permeases are extremely selective enzymes, and if the microorganism lacks the permease for a particular substrate, it will not be able to utilise that substrate for growth even when it is present at high concentrations. This observation helps to explain why many apparently slow-growing bacteria are able to establish stable populations in wastewater communities. Frequently they are the sole organisms which possess the necessary carrier to transport a particular component of the wastewater. By means of active transport, bacteria are able to maintain internal substrate concentrations up to a thousandfold higher than the level in the external environment. This permits them to grow in a wide range of habitats such as rivers and oceans, where the BOD is so low as to be unmeasurable.

5.4 OXIDATION OF ORGANIC SUBSTRATES

Nutritional classifications

Microorganisms are capable of utilising a wide range of substrates to provide the essential elements of their nutrition. A simple classification based on their source of energy, carbon and reducing equivalents is used to categorise their nutritional requirements. Organisms which are capable of photosynthesis derive their energy directly from sunlight and are known as phototrophs, the remainder which generate energy from oxidation of organic or inorganic compounds are known a chemotrophs. Certain organisms are capable of fulfilling all their requirements for cell carbon from simple inorganic carbon compounds such as CO_2, bicarbonate and carbonate and are known as autotrophs; heterotrophs, however, require a fixed source of organic carbon. Finally, many organisms are capable of producing the reducing equivalents necessary for energy generation from oxidation of inorganic compounds such as ferrous iron (Fe^{2+}) or ammonium (NH_4^+) and these are known as lithotrophs. Organisms which obtain their reducing equivalents from oxidation of organic compounds are known as organotrophs. The terms 'heterotroph' and 'autotroph' are used widely by engineers, and organisms which are responsible for the removal of BOD are lumped under the term 'heterotroph'. Similarly, organisms which carry out ammonia removal are termed autotrophs. However, these terms are not popular with microbiologists, as in the case of autotrophs and lithotrophs it is difficult to distinguish between inorganic sources of carbon and of reducing equivalents. Consequently the four terms: photolithotroph, photoorganotroph, chemolithotroph and chemoorganotroph prove adequate to describe all the nutritional types found in wastewater treatment.

Important catabolic pathways

Chemoorganotrophs are capable of oxidising a large array of carbon substrates as carbon and energy sources. The majority of organisms have a few preferred substrates such as certain carbohydrates and amino acids. Other organisms display a wide versatility and can oxidise not only the waste products of mammalian metabolism but also many synthetic organic compounds which are the end-products of industrial processes. Despite these widely differing capabilities, the means by which these oxidations are achieved are remarkably similar in all living organisms. The ultimate aim of energy-generating reactions is to provide energy for the synthesis of ATP. However, organic compounds possess energy which is well in excess of this requirement. For instance, the complete oxidation of glucose would provide 2872 kJ of energy and if this was used with 100% efficiency it would be enough for the synthesis of 80 molecules of ATP.

$$C_6H_{12}O_6 + 6O_2 \longrightarrow 6CO_2 + 6H_2O \quad \Delta G = -2872 \text{ kJ/mol}$$

(5.16)

If all this energy was released at once it would prove disastrous for the microorganism. Oxidation proceeds therefore in a series of reactions catalysed by enzymes, which are organised to release energy in discrete and controllable amounts thus enabling them to be coupled to ATP synthesis. Such a series of reactions is known as a metabolic pathways, and the same major pathways are exploited by all organisms. A second advantage of such a mechanism of degradation is that a large number of different carbon skeletons are produced which can be used as building blocks for the synthesis of

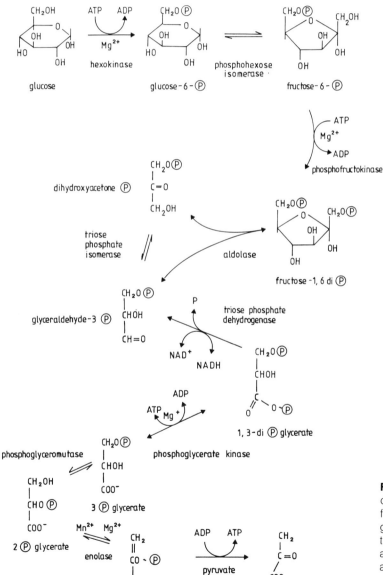

Figure 5.12 Glycolysis or the EMP pathway for the degradation of glucose to pyruvate in the presence or absence of oxygen, associated with the synthesis of two molecules of ATP.

new cell material by the process of anabolism. Research into metabolic pathways has traditionally been carried out using glucose as the substrate, and this convention has been followed here although glucose is a minor component of the BOD of most wastewaters. The mode by which other compounds such as amino acids and fats are oxidised and enter these pathways will be described where appropriate.

After the transport of an organic substrate across the cell membrane, a microorganism will convert this in a series of steps to a three-carbon compound known as pyruvic acid. Microorganisms can exploit one (or more) of four possible metabolic pathways to achieve this aim. These pathways differ in their initial reactions but share a

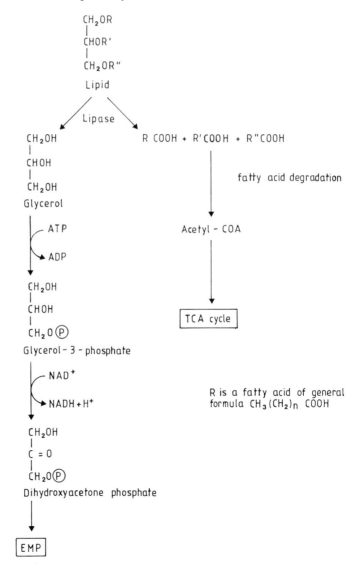

Figure 5.13 The degradation of lipids to small molecules, and their entry into the major catabolic pathways.

common mechanism for the conversion of the phosphorylated three-carbon sugar, glyceraldehyde 3-phosphate, to pyruvate. Two of these four pathways are discussed here in detail in order to illustrate the general features of metabolic pathways.

The most widely used mechanism is known as glycolysis or the Embden–Meyerhof–Parnas (EMP) pathway (Figure 5.12). The pathway has three main stages: the phosphorylation of the six-carbon sugar glucose using phosphate from ATP; the cleavage of this six-carbon sugar into two three-carbon phosphorylated sugars; and finally the cleavage of the phosphoryl bond to generate ATP and yield pyruvate. In the early stages of degradation, two molecules of ATP are required to form the high-energy compound glyceraldehyde 3-phosphate, and two molecules of this three-carbon sugar are produced for each molecule of glucose oxidised. Glyceraldehyde 3-phosphate is the linkage point of glycolysis with fat and lipid degradation. Lipids are initially broken down to produce glycerol and fatty acids by means of the enzyme lipase. Glycerol is phosphorylated and oxidised to dihydroxy acetone phosphate and fatty acids are oxidised to acetate (Figure 5.13). The initial energy expenditure of two ATP molecules is now repaid by two substrate-level phosphorylation reactions which result in the synthesis of four molecules of ATP. The net result of glycolysis, therefore, is the production of two ATP molecules and two pyruvate molecules for each glucose molecule degraded. This has been at the expense of NAD^+ which is reduced to NADH.

A second pathway known as the Entner–Doudoroff (ED) pathway, is less widely utilised than glycolysis but is exploited by members of the genus *Pseudomonas* and certain enteric bacteria. The pseudomonads are a very versatile genus which can oxidise large numbers of different compounds including hydrocarbons, phenols and aliphatic amides. They are universal in wastewater treatment reactors and it is probable that they play a major role in BOD removal. The ED pathway has end-products of pyruvate and glyceraldehyde 3-phosphate which may be converted to a second pyruvate molecule by reactions common to glycolysis (Figure 5.14). Under anaerobic conditions the ED pathway produces only 1 mole of ATP for each mole of glucose oxidised. This occurs because only 1 mole of triose phosphate is produced and oxidised as compared with the two in glycolysis.

Of the other two sequences, the pentose phosphate pathway is important because reducing equivalents are donated to $NADP^+$ and not NAD^+ and this is required in the reductive step of many anabolic reactions. This pathway lacks the enzyme which cleaves glucose phosphate into two three-carbon sugars but instead produces a five-carbon sugar (or pentose). This is split into a two-carbon fragment and triose phosphate by the enzyme phosphoketolase. Only one molecule of ATP per molecule of glucose is produced. The final

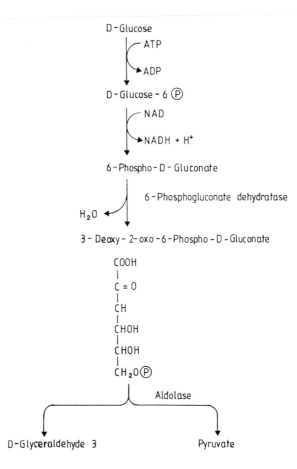

Figure 5.14 The ED pathway for glucose catabolism associated with the production of one molecule of ATP under anaerobic conditions.

sequence is known as the phosphoketolase pathway. This has the first five reactions in common with the pentose pathway except that NAD^+ and not $NADP^+$ is used. The end-products are lactate, ethanol and carbon dioxide and only 1 mole of ATP is produced per mole of glucose oxidised.

The above reactions have no requirements for oxygen and will thus occur under both aerobic and anaerobic conditions. However, before catabolism can proceed further the microorganism must regenerate its supply of NAD^+ which has been reduced during the oxidation of glucose. Two mechanisms are available whereby an organism can achieve this. One of these is dependent upon the presence of oxygen, the other can occur under anaerobic conditions.

Respiration

The role of the electron transport chain in the oxidation of NAD(P)H has been discussed previously. Organisms which possess a functional electron transport chain are therefore able to regenerate NAD^+ by this mechanism in the presence of oxygen with each molecule of NADH oxidised, being coupled to the production of three molecules of ATP. In the absence of oxygen, certain organisms are able to exploit other terminal electron acceptors. Of particular importance is the utilisation of nitrate (NO_3^-) as a terminal electron acceptor with the production of nitrate gas, a process known as denitrification. This occurs in the absence of oxygen and is widely used in wastewater treatment to remove nitrogen from an effluent. When a water lacks gaseous dissolved oxygen but contains oxygen which is accessible in the form of nitrate or sulphate, it is frequently termed anoxic, to differentiate it from anaerobic conditions, under which oxygen is not available in any form. If NADH is oxidised by respiration then pyruvate can be completely oxidised to carbon dioxide and water by means of the tricarboxylic acid (TCA) cycle which is often referred after its discoverer as the Krebs cycle. The TCA cycle, as its name

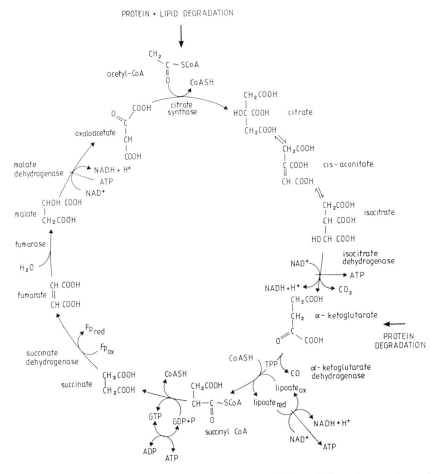

Figure 5.15 The TCA or Krebs cycle for the aerobic degradation of pyruvate and associated small molecules under aerobic conditions.

suggests, is a cyclical pathway which allows access for the end-products of other pathways, such as amino-acid and fat degradation, at several points (Figure 5.15). Similarly the intermediates which are produced in this cycle may be removed to be used as carbon skeletons in anabolism. In addition to this role, the TCA cycle is also the major energy-generating mechanism under aerobic conditions. NAD^+ is the acceptor of reducing equivalents in four reactions which occur during the TCA cycle and upon reoxidation via the electron transport chain, a total of 12 molecules of ATP will be produced for each molecule of pyruvate (or 24 molecules of ATP per molecule of glucose). A further 3 molecules of ATP are produced from the reactions listed in Table 5.2 and thus the complete aerobic oxidation of glucose by glycolysis and the TCA cycle, results in the production of 38 molecules of ATP.

Fermentation

Fermentation allows a microorganism to reoxidise NAD(P)H in the absence of oxygen by utilising an organic compound as an electron

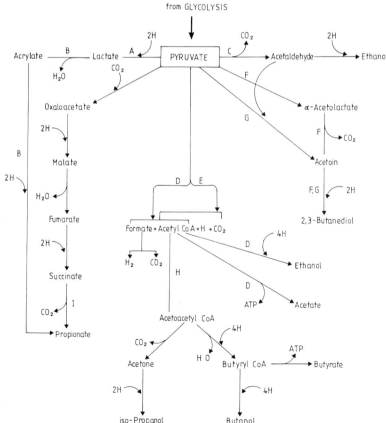

Figure 5.16 Pathways exploited by microorganisms for the reoxidation of NAD(P)H in the absence of oxygen. The highly reduced end-products are associated with a particular genera of microorganisms. (From Mandelstam *et al.* Reproduced by permission of Blackwell Scientific Publications Ltd.)

acceptor. NAD(P)H loses its hydrogen atoms in a reaction coupled to the reduction of pyruvate and the product of this reduction reaction is the end-product of glucose catabolism. Several such highly reduced end-products are produced and excreted into the surrounding medium, each one characteristic of a particular organism (Figure 5.16). A feature of many of these compounds is their highly objectionable smell. Thus if a wastewater or sludge receives inadequate oxygenation, then the component organisms are forced to carry out fermentative catabolism. This is the primary reason for odours at sewage treatment works and the problem is rapidly alleviated by aeration. A group of organisms known as methanogens are capable of coupling the oxidation of these products with the reduction of carbon dioxide to methane, in a process known as anaerobic digestion. Methanogens are fastidious bacteria which only proliferate under controlled conditions, and fermentation in the absence of methanogens is frequently termed putrefaction. Because fermentation ends with the excretion of highly reduced end-products, the energy still remaining in these compounds is lost to the organism. Fermentation is a very inefficient process compared to respiration since no further ATP is produced by these reduction reactions. Anaerobic growth processes, therefore, have very low yields (g sludge produced/g BOD degraded) and this feature is exploited in anaerobic digestion in order to reduce the quantity of sludge produced during aerobic treatment.

5.5 OXIDATION OF INORGANIC SUBSTRATES

Microorganisms which obtain their energy from the oxidation of inorganic compounds are termed chemolithotrophs. Bacteria which demonstrate this mode of energy generation couple the energy release to ATP biosynthesis by means of the electron transport chain. In addition to their unusual energy sources, chemolithotrophs are frequently autotrophic, generating all their cell carbon requirements from carbon dioxide. Carbon dioxide is assimilated into the cell by means of the Calvin cycle which is discussed below. However, some chemolithotrophs are capable of heterotrophic growth and some practice both heterotrophy and autotrophy (sometimes referred to as mixotrophy), depending upon the nutrient availability. These are a large number of potential inorganic sources of energy which includes hydrogen, ammonia, many metal ions and sulphur. Of particular importance in wastewater treatment are the oxidation of ammonia, sulphur and ferrous iron (Fe^{2+}).

Ammonia oxidation

Ammonia is oxidised ultimately to nitrate in two reactions carried out by two distinct groups of obligately aerobic bacteria. The first intermediate is nitrite and this reaction is catalysed by the genus *Nitrosomonas*.

$$NH_3 + 1.5O_2 \longrightarrow NO_2^- + H^+ + H_2O \qquad \Delta G^{0\prime} = -272\,kJ/mol$$

(5.17)

This energy is coupled to ATP synthesis only at site 3 of the electron transport chain and thus only 1 mol of ATP is produced per mole ammonia oxidised.

Nitrite is further oxidised to nitrate by *Nitrobacter*. The oxygen atom incorporated into the nitrate does not derive from molecular oxygen as in the oxidation of ammonia, but from water. This was clearly shown from experiments in which water containing a radioisotope of oxygen (^{18}O) was employed. The nitrate thus synthesised, was found to have incorporated this oxygen isotope:

$$NO_2^- + H_2{}^{18}O + 0.5O_2 \longrightarrow N^{18}O_3^- + H_2O$$
$$\Delta G^{0\prime} = -74.8\,kJ/mol \qquad (5.18)$$

Nitrosomonas and *Nitrobacter* are both autotrophic genera which assimilate carbon dioxide as a source of carbon via the Calvin cycle. The energy expenditure required to achieve this is very high, and because of the low energy yield resulting from ammonia and nitrite oxidation these organisms demonstrate very low growth yields.

Iron oxidation

Iron is oxidised by the iron bacteria according to the reaction:

$$Fe^{2+} \longrightarrow Fe^{3+} + e^- \qquad (5.19)$$

The resultant ferric salt gives these organisms their typical brown coloration. As the redox potential of the Fe^{2+}/Fe^{3+} couple is $+0.772\,V$, it is difficult to see how coupling this to the electron transport chain can yield sufficient energy to drive ATP synthesis. It has been suggested, therefore, that iron oxidation is not an essential feature of their energy generation. Certain iron bacteria have an optimal pH for growth of around 2 which produces a pH differential of 4.5 units across the membrane. This will contribute 270 mV to the proton motive force and may help to explain their energy source. Over fifteen genera of bacteria are able to oxidise Fe^{2+} and those of importance in wastewater treatment include *Sphaerotilus*, *Gallionella* and *Leptothrix*. These are generally associated with problems in water supplies. Typical symptoms include: discoloration of water (blood-red or brown); reduction in flow rates due to iron bacteria coating the pipes in the distribution system; clogging of filter screens; and the development of this red coating in reservoirs, tanks and cisterns.

Sulphur oxidation

Many bacteria can obtain their energy from the oxidation of sulphide, sulphur, thiosulphate and other reduced sulphur compounds to sulphate. Consequently they are an important component of the global sulphur cycle. *Thiobacillus* is the most widespread genus of sulphur oxidisers. In addition the filamentous bacterium *Beggiatoa* is capable of oxidising sulphide and accumulating elemental sulphur which can be further oxidised to sulphate if the external sulphide concentration is low. The reaction of most importance in wastewater treatment is the oxidation of hydrogen sulphide with the production of sulphuric acid according to the reaction:

$$H_2S + 2O_2 \longrightarrow H_2SO_4 \qquad (5.20)$$

Hydrogen sulphide is frequently produced in sewer pipes by the action of sulphate reducers when the wastewater becomes anaerobic. This is given off into the atmosphere where it is oxidised by the sulphur oxidisers which can attach to the walls of the sewer pipes (Figure 5.17). The sulphuric acid which is produced will attack both cast-iron and concrete sewer pipes and this is frequently severe enough to cause collapse. The bacterium *Thiobacillus ferrooxidans* is often implicated in such attacks as it does not require gaseous hydrogen sulphide, being capable of oxidising both insoluble sulphides and iron. The end-product of these reactions are ferric hydroxide and sulphuric acid and *T. ferrooxidans* is quite capable of growing at pH values as low as 1, which result from the production of sulphuric acid. The reaction starts with the oxidation of sulphide to sulphate:

$$2FeS_2 + 7O_2 + 2H_2O \longrightarrow 2FeSO_4 + 2H_2SO_4 \qquad (5.21)$$

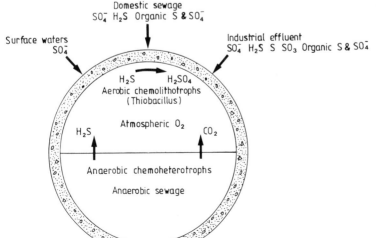

Figure 5.17 The inputs of sulphur to a sewerage system and its oxidation to sulphuric acid, which can result in severe corrosion.

ferrous iron is then oxidised to ferric ion:

$$4FeSO_4 + 2H_2SO_4 + O_2 \longrightarrow 2Fe_2(SO_4)_3 + 2H_2O \quad (5.22)$$

and this results in the precipitation of ferric hydroxide:

$$Fe_2(SO_4)_3 + 6H_2O \longrightarrow 2Fe(OH)_3 + 3H_2SO_4 \quad (5.23)$$

Autotrophic carbon assimilation

Both eukaryotic and prokaryotic autotrophs employ the same metabolic pathway for the incorporation of carbon dioxide into cell carbon, a process termed 'fixation'. This pathway known as the Calvin cycle is closely associated with the pentose phosphate pathway for glucose degradation. Carbon dioxide is reduced by reaction with a pentose sugar, ribulose 1,5-diphosphate, to yield two molecules of the triose sugar, 3-phosphoglycerate:

$$\text{Ribulose 1,5-diphosphate} + CO_2 \longrightarrow \text{2,3-phosphoglycerate}$$
$$(5.24)$$

Ribulose 1,5-bisphosphate arises directly from the pentose phosphate pathway by phosphorylation of the intermediate ribulose 5-phosphate:

$$\text{Ribulose 5-phosphate} + ATP \longrightarrow \text{ribulose 1,5-bisphosphate} + ATP$$
$$(5.25)$$

The relationship of the Calvin cycle to the pentose phosphate pathway is shown in Figure 5.18. The net result of one complete cycle is

$$6CO_2 + 6 \text{ ribulose 1,5-bisphosphate} + 18ATP + 12NAD(P)H$$
$$+ 12H^+ + 2H_2O \longrightarrow \text{fructose 6-phosphate}$$
$$+ 6\text{-ribulose 1,5-bisphosphate}$$
$$+ 18ADP + 17Pi + 12NAD(P)^+ \quad (5.26)$$

Figure 5.18 The association of the pentose phosphate pathway and the Calvin cycle for incorporation of carbon dioxide into organic carbon, exploited by autotrophic microorganisms.

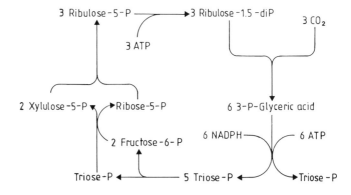

It is apparent that this pathway has a high-energy demand as it requires eighteen molecules of ATP and twelve molecules of NAD(P)H to synthesise a single molecule of hexose sugar. This may represent 80% or more of the energy requirements of autotrophs. Autotrophs, therefore, are poor competitors with heterotrophs in wastewaters where an adequate supply of organic carbon is required. When this carbon is depleted and oxygen is still available, this situation is reversed as the heterotrophs will lack both a carbon and an energy source. Such a situation is engineered in wastewater treatment plants in order to achieve ammonia removal by nitrification.

6 The Kinetics of Bacterial Growth

6.1 INTRODUCTION

When placed in a suitable growth medium and under favourable environmental conditions of pH, temperature and oxygen concentration, microorganisms will proceed to assimilate nutrients, leading to an increase in the size of the population as a result of growth and reproduction. The rate at which these substrates are assimilated and the biomass in the system accumulates, is reproducible for a given set of conditions and may be described by simple mathematical models. These models are of use both as an aid to the fundamental understanding of the mechanisms which influence growth and, as will be seen later, in providing a framework for the design and operation of wastewater treatment systems.

Because of the ease in which microorganisms may be grown in pure cultures on defined media, and because of the large number of reliable and rapid techniques available for the quantification of growth and substrate removal, much research has been carried out in this area. This has resulted in reliable kinetic models and equations to describe the growth process. These simple models form the basis for many of the more complex models used to describe the growth of mixed microbial cultures on wastewaters. As such they must be fully understood in order to appreciate their uses and limitations when used in this way.

Determination of microbial growth

A direct result of microbial growth is an increase in both the number of individuals (N) in the population and the weight, or biomass (X), of the population. Thus any technique which measures these changes accurately may be used to assess the extent of microbial growth. The easiest technique therefore is simply to weigh the biomass directly, after removing it from the bulk liquor, washing it to remove residual liquor and drying to a constant weight. The cells may be separated from the liquor either by centrifugation, or filtration through a filter with an appropriate pore size (generally $0.7\,\mu m$).

This latter is the method of choice that should be used with wastewater samples, although an alternative and more rapid technique is to correlate the turbidity exhibited by a suspension of microorganisms with its biomass concentration. Dispersed suspensions of microorganisms behave like a colloidal suspension and will scatter a monochromatic beam of light passed through the suspension. The degree of light scattering is a function of the number of cells in suspension, their shape and other factors such as their possession of refractile granules (e.g. phb or polyphosphates). This light scattering

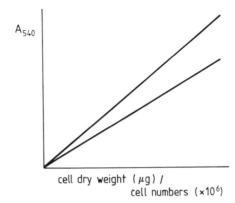

Figure 6.1 The relationship between the weight or number of a bacterial population and its turbidity, measured as absorption at a wavelength of 540 nm.

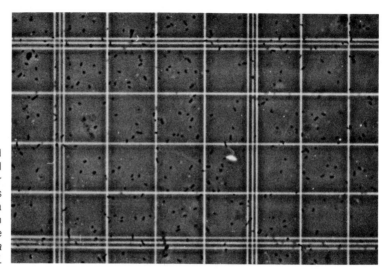

Figure 6.2 The ruled grids of a bacterial counting cell, or haemocytometer, as seen through a microscope, when counting the bacterium *Escherichia coli.*

may be quantified either by measuring the amount of light which is scattered at 90° to the light beam using a turbidimeter, or measuring the total amount of light scattered as a difference in absorbance when the suspension is placed in a spectrophotometer. The results allow the construction of a calibration curve which correlates either turbidity or absorbance, with biomass concentration (Figure 6.1). This calibration curve is very specific for a given organism and the application of the technique to heterogeneous populations, as would be found in wastewater samples, is not recommended particularly when such phenomena as microbial flocculation, make determination of light scattering difficult. It is, however, ideal for pure culture studies on dispersed populations.

In addition to preparing calibration curves which relate dry weight to absorbancy, similar curves can be prepared which relate cell numbers to either absorbancy or dry weight. Cell numbers are generally determined by direct microscopic counting utilising a special cell, known as a haemocytometer, which holds a known volume of liquid and is divided into ruled grids to facilitate counting (Figure 6.2). Again this technique is not recommended for wastewater samples which, owing to their flocculent nature, make accurate counting difficult.

6.2 GROWTH OF MICROORGANISMS IN BATCH CULTURE

This section will focus primarily on the growth of pure cultures of microorganisms on simple defined media, in which there is a single source of carbon which is also the energy source. Growth of the microorganism is determined by changes in the weight of biomass, and changes in substrate concentration (S) are determined using a specific, sensitive assay for the substrate. The majority of pure culture studies have used glucose as the substrate.

When grown in batch culture, the change in microbial population with time follows the classical growth curve illustrated in Figure 6.3. This curve has three quite distinct phases, namely lag, exponential and stationary.

The lag phase of growth represents the acclimation period of an organism to its new environment and is of interest to wastewater engineers from the point of view of process start-up. It is also of profound importance in the BOD_5 test. Although biomass is not changing during this period, a lot of metabolic activity is taking place within the organism, and substrate is being assimilated and utilised for the synthesis of new enzymes and for growth of the cell prior to division. The length of the lag period is very variable and reflects both the metabolic state of the cell on transference to its new environment, and also differences in the substrate composition between its present and previous environment, which may require synthesis of several

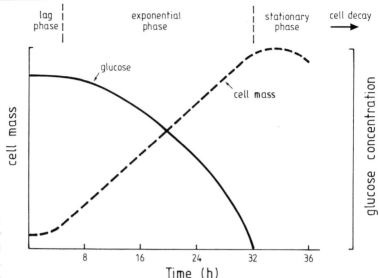

Figure 6.3 The growth curve demonstrated by a typical bacterial species growing in a liquid culture medium.

new enzymes before division takes place. Microorganisms transferred during the exponential phase of growth from one medium to another one of identical composition generally demonstrate no lag period; conversely, an organism taken from late stationary phase of growth, transferred to a medium containing a different carbon source and at a different temperature, may exhibit a lag period of several days (and indeed may not survive at all!).

On commencement of cell division the organism moves out of the lag period and continues dividing exponentially until such time as the medium is no longer able to support growth. There are many reasons for this of which the most common are: depletion of a nutrient essential for growth; depletion of the dissolved oxygen supply; excretion of toxic end-products of metabolism; and changes in medium pH owing to the excretion of acidic end-products of metabolism. The growth curve illustrated in Figure 6.3 shows growth ceasing as a result of glucose depletion. This exponential phase of growth is the most important part of the growth curve as it represents the maximum rate of substrate removal, and in wastewater treatment the aim is to remove substrate as rapidly as possible.

It has been found that, for the exponential phase of growth, the rate of increase in biomass is directly proportional to the initial cell concentration and it may be represented by a first-order reaction of the form:

$$\frac{dX}{dt} = \mu X \tag{6.1}$$

Where the first-order rate constant (μ) is known as the specific growth rate, with units of mass cells produced/mass cells present $\times t$, which reduces to t^{-1}.

If the cell concentration at t_0 is X_0, and after an interval (t) has increased to a new concentration X_t, equation (6.1) can be integrated between the limits t and t_0 to yield the equation of a first-order reaction:

$$X_t = X_0 e^{\mu t} \tag{6.2}$$

and this may be used to calculate the weight of cells at any given time from a knowledge of the initial cell concentration and the specific growth rate. The amount of time taken for a cell population to double in weight can also be calculated from this equation by letting the cell population equal twice the original population (i.e. $X = 2X_0$) for a time interval of t_d, where t_d is the doubling time, thus:

$$2X_0 = X_0 \exp(\mu t_d) \qquad 2 = \exp(\mu t_d) \qquad \ln 2 = \mu t_d \tag{6.3}$$

The specific growth rate (and hence the doubling time) of a given microorganism growing on a given medium is calculated graphically from results of batch growth experiments which determine the increase in biomass with time. Taking logs of equation (6.2) shows that a plot of ln X against time will be linear over the exponential period with a gradient of μ:

$$\ln X = \ln X_0 + \mu t \tag{6.4}$$

It is generally more convenient, however, to plot the data directly on to semilogarithmic graph paper in which case the gradient will equal $\mu/2.303$.

If we now consider what is happening to the concentration of substrate during the growth of the organism, a similar set of equations may be constructed for substrate utilisation. Figure 6.3 illustrates that glucose is utilised throughout the growth curve until after 32 hours it becomes exhausted. It is at this point that the organism enters the stationary phase of growth and it is reasonable to assume that this is a direct result of substrate depletion. In this case glucose is said to be

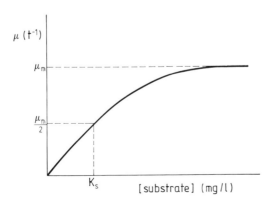

Figure 6.4 The relationship between the growth rate of a microorganism and the concentration of growth-limiting substrate in the culture medium.

the growth-limiting substrate. The depletion of substrate also follows first-order removal kinetics during the exponential phase, with removal rate being proportional to the number of cells present. The first-order removal constant is known as the substrate utilisation rate (q) and has units of g substrate removed/g cells $\times t$, or more usually t^{-1}.

$$\frac{ds}{dt} = q \cdot X \qquad (6.5)$$

It is often very useful, when formulating models of substrate removal during wastewater treatment, to know what mass of cells will be produced from a given amount of substrate. This may be found from Figure 6.3 by determining the amount of cells produced over a given increment of time and then calculating how much glucose has been utilised over the same time period. The mass of cells produced per unit of substrate removal is known as the cell yield (Y), where

$$Y = \frac{dX/dt}{ds/dt} \qquad (6.6)$$

Although the value for Y will vary depending on the nature of the substrate, the microbial species and the environmental conditions under which the cell is grown, it is assumed constant for a given set of experimental conditions and also for heterogeneous populations growing on a specific wastewater (for example domestic wastewater).

There are many factors which influence the magnitude of the yield coefficient of which the most important is the oxidation state of the carbon source. The oxidation state of bacterial cell material is approximately the same as that of carbohydrate, consequently if the organism is growing on a carbon source which is at a higher oxidation state than this, then expenditure of energy will be required to reduce it to the required oxidation state. Conversely, if the carbon source is more reduced it will be oxidised to the proper level during normal oxidation and no extra energy expenditure will be required. A similar argument applies to the number of carbon atoms in the carbon source. It has been shown that three carbon atom fragments (such as pyruvic acid and 3-phosphoglycerol) play a central role in metabolism, and carbon sources which have less than three carbon atoms will require the expenditure of energy for synthesis. Consequently, substrates containing few carbon atoms require more energy for synthesis than do large ones. These findings help to explain the low yield coefficients demonstrated by the autotrophs. These organisms utilise CO_2 as their sole source of carbon and this compound is in the highest possible oxidation state for carbon, as well as being a one-carbon compound. The energy requirements for synthesis are therefore very high, and the amount of cell material which can be formed per unit of charged energy carrier is low.

Other factors likely to affect the yield coefficient include: the

Bacteria	k_d (d^{-1})	μ_m (d^{-1})	K_s (mg/l)	Y (mg cells/ mg substrate)	M (d^{-1})
Zooglea ramigera	0.08	5.5	0.3	0.51	0.16
Sphaerotilus natens	0.05	6.5	10.0	0.53	—
Haliscomenobacter hydrossis	0.06	1.2	8.0	0.59	0.5
Escherichia coli	0.06	25.2	15.0	0.66	1.4
Type 0092	0.07	7.5	350	0.55	0.12

Table 6.1 Kinetic coefficients of bacteria associated with wastewater treatment processes.

formation of storage products such as glycogen, phb and polyphosphate; changes in the viable fraction of microorganisms; and changes in the portion of substrate which is used for maintenance energy. Many of these are likely to be of importance in wastewater treatment processes and therefore appropriate factors must be incorporated into mathematical models in order to correct for them. Alternatively, the experimental conditions under which numerical values for the yield are determined must be similar to those which will be encountered in practice.

Typical heterotrophic microorganisms have values for the yield coefficient in the range 0.4–0.6 g biomass/g substrate (Table 6.1), which indicates that up to 60% of the substrate utilised has been used for purposes other than biomass production.

Equation (6.6) may now be rearranged by the substitution of equations (6.1) and (6.5) to produce a useful derivative which allows a direct calculation of the yield from a knowledge of the growth rate of the organism and the rate at which it utilises substrate:

$$Y = \frac{\mu}{q} \tag{6.7}$$

Effect of substrate concentration on microbial growth

In a closed environment such as a batch culture, the decline in bacterial growth rate observed at the end of exponential growth is generally related to the depletion of a single nutrient known as the growth-limiting nutrient. The growth-limiting nutrient could be the organic carbon source, nitrogen, phosphorus or any other of the many factors needed for bacterial growth. In addition the magnitude of the specific growth rate exhibited by an organism is dependent upon the concentration of the growth-limited nutrient. This phenomena was first recognised and described by Jacques Monod in 1949 from experiments performed in batch culture. The determination of the concentration of growth-limiting nutrient from batch culture experiments poses many analytical difficulties because it is generally very low at the cessation of exponential growth, frequently below the limits of reliable detection. Monod, however, was able to analyse data mathematically from batch growth curves using the kinetic equations derived earlier. The value for the yield coefficient was calculated from the growth curve using the difference in cell and substrate concentration before and after exponential growth. In addition the term dX/dt was evaluated at several points, from the gradient of tangents drawn at points to the growth curve during the transition from the exponential to stationary phase of growth. These values permitted calculation of the organism's specific growth rate using equation (6.1). In addition, knowing the yield coefficient, the residual substrate concentration could be calculated for each of the points plotted along the growth curve by application of equation (6.6).

The resulting curve was similar to that depicted in Figure 6.4 and this observation has been substantiated many times over the last 40 years. Insufficient information was available to Monod regarding the mechanisms of microbial growth to permit him to propose a mechanistic equation to describe his observations (and indeed is not available to this day). However, he was able to propose an empirical relationship which showed that the curve could be approximated adequately by the equation for a rectangular hyperbola (which is also analogous to the Michaelis–Menton equation which describes enzyme reaction kinetics), namely:

$$\mu = \mu_m \frac{S}{K_s + S} \tag{6.8}$$

where μ_m is the maximum specific growth rate (with units of t^{-1} and it is the growth rate which would be observed in the absence of any substrate limitation; K_s is a saturation coefficient with units of mg/l, defined as the growth-limiting substrate concentration which allows the organism to grow at half the maximum specific growth rate; S is the concentration of the growth limiting substrate with the same units as K_s. Thus μ_m represents the asymptote to a rectangular hyperbola and K_s is the shape factor.

The saturation coefficient is a very important parameter in describing the outcome of competition between different microorganisms competing for a limited supply of food. It is a measure of the affinity an organism has for the growth-limiting nutrient; the lower the value for K_s then the greater is the organisms affinity for substrate. This suggests that in an environment with low growth-limiting substrate concentration (such as a wastewater treatment process), organisms demonstrating the lowest saturation coefficients will possess a greater capacity to grow rapidly. The influence of K_s and μ_m on the growth of two organisms competing for a single limiting substrate is illustrated in Figure 6.5. The outcome is decided by μ_m if the food supply in excess, with the organism with the higher μ_m outgrowing organisms with lower values of μ_m. Conversely, at limiting substrate concentrations, an organism with a lower satur-

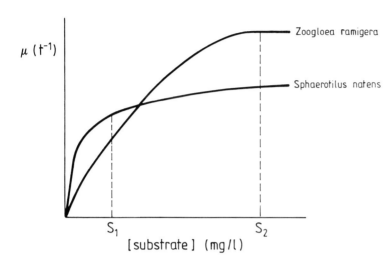

Figure 6.5 The effect of the concentration of growth-limiting substrate on the outcome of competition between two organisms with different values of μ_m and K_s. At a low substrate concentration (S_1), *Sphaerotilus natans* will be the predominant organism due to its higher substrate affinity (lower K_s). At a high substrate concentration (S_2), *Zoogloea ramigera* will predominate as it has a higher maximum specific growth rate (μ_m).

ation coefficient will prove a more successful competitor.

If the Monod equation (6.8) is now substituted for the specific growth-rate term in equation (6.1), an expression results which allows us to predict the kinetics of substrate removal for microorganisms growing at different substrate concentrations, assuming their saturation coefficient is known:

$$\frac{dX}{dt} = \frac{\mu_m S}{K_s + S} X \qquad (6.9)$$

This equation shows that microorganisms demonstrate growth rates which are first order with respect to cell concentration and either zero or first order with respect to substrate concentration. For a microorganism growing in an environment in which the substrate concentration is high compared to K_s, then the substrate term disappears from equation (6.9) showing that the growth rate of the organism is zero order with respect to substrate concentration:

$$\frac{dX}{dt} = \mu_m X \qquad (6.10)$$

If, however, substrate concentration is low compared to K_s, then the S term in the denominator becomes unimportant and equation (6.9) may be approximated as a first-order reaction with respect to substrate removal:

$$\frac{dX}{dt} = \frac{\mu_m S}{K_s} X \qquad (6.11)$$

Activated sludge systems are characterised by low substrate concentrations and populated with organisms which display a high saturation coefficient, thus equation (6.11) may be used to describe these systems and there is much experimental evidence to justify such a choice.

6.3 MICROBIAL GROWTH IN CONTINUOUS CULTURE

There are many examples of the industrial exploitation of batch culture techniques of which the brewing industry is perhaps the best known. However, this technique for bacterial growth proves very inefficient, as it involves a long unproductive period at the cessation of growth during which the product is harvested, the apparatus replenished with medium, inoculated with the appropriate micro-organism and the process repeated. Clearly this unproductive period can be eliminated from the process by operating the system cont-inuously. Indeed this mode of operation has been practised at wastewater treatment works for over 60 years, but it is only within the last 20 years that the potential of continuous culture has attracted much attention within industry.

The usage of continuous-culture techniques was pioneered largely by researchers in the area of microbial physiology, who were limited by the constraints of batch growth experiments. During microbial growth in batch culture, bacteria are faced with a constantly changing environment in terms of medium composition, pH, oxygen con-centration, population density and many other factors. Consequently their physiology will be constantly changing in order to adapt to the changing environment, and this makes the interpretation of the effects of environmental conditions on microbial physiology very difficult. Continuous-culture techniques make it possible to maintain a growing culture under steady-state conditions at a defined and constant growth rate, when such factors as medium composition, pH, temperature and dissolved oxygen concentration are all held const-ant. Thus continuous culture can be used to maintain cells at constant defined growth conditions with a constant physiology, and to study the response of their physiology to perturbations in a single environmental parameter. Results from such studies have had widespread industrial applications and the technique has played a prominent part in the so-called 'biotechnology revolution'.

All continuous-culture apparatus comprises three basic units, namely:

1. A reservoir to hold the bacterial growth medium.

2. A reactor or culture vessel, which receives medium from the reservoir and where the organism may be grown in a controlled environment, under defined conditions.

3. A reservoir to hold the spent medium and cells which emerge from the reactor.

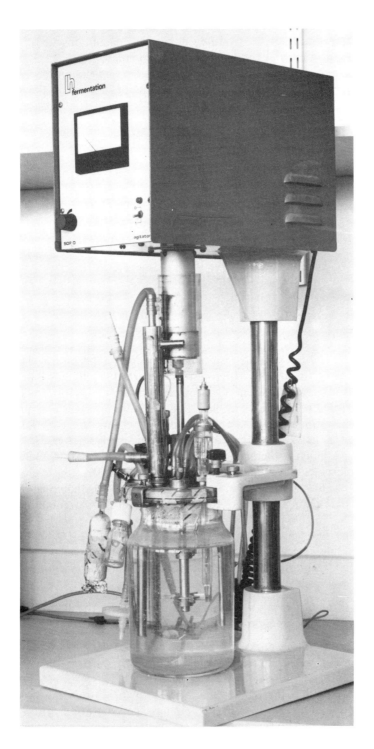

Figure 6.6 A
laboratory-scale
continuous-culture
apparatus (chemostat)
of 1 l operating volume.

In addition a suitable pump is required to deliver medium from the reservoir to the reactor at a constant but variable flow rate, and the reactor should possess an outflow device which maintains the volume constant. Such apparatus is available commercially from a large number of manufacturers in sizes ranging from small laboratory-scale reactors of 500 ml up to pilot-plant facilities of 200 m^3 and above. A typical commercially available laboratory set-up is illustrated in Figure 6.6.

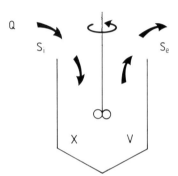

Figure 6.7 Schematic diagram of an idealised continuous-culture system.

Let us now consider an idealised continuous-culture system (Figure 6.7) in which a stirred vessel of volume V containing biomass at a cell concentration X and a growth-limiting substrate concentration of S_e, is fed at a flow rate Q from a reservoir containing medium with a growth-limiting substrate concentration of S_i. Such a vessel is known as a continuous-flow, stirred-tank reactor (CFSTR), and by applying the principles of mass balance around this reactor simple mathematical models can be constructed to describe microbial growth and substrate removal. The following reasonable assumptions are required in order to formulate the necessary equations:

1. The inflowing medium is instantly dispersed throughout the culture vessel (i.e. the mixing regime follows complete-mix kinetics). This implies that the biomass and substrate concentration in the effluent will be the same as those in the reactor.

2. All the organisms are distributed randomly in the culture vessel with no clumping or wall growth occurring.

3. A constant amount of biomass is produced per unit of growth-limiting substrate (i.e. the yield is constant).

Thus within the reactor microbial growth is occurring, with a corresponding increase in the biomass concentration. However, concomitant with this, biomass is also being lost from the system as a result of culture overflow and biomass balance in the system results which may be represented as

Rate of change of   Rate of biomass   Rate of
biomass concentration = production   − biomass   (6.12)
in the system   removal

Expressing this mathematically for a unit volume of culture V, where bacterial growth is represented by equation (6.1):

$$\frac{dX}{dt} V = \mu X V - Q X \qquad (6.13)$$

dividing by V:

$$\frac{dX}{dt} = \mu X - \frac{Q X}{V} \qquad (6.14)$$

If this reactor is now operated under steady-state conditions (which implies that $dX/dt = 0$), then:

$$\mu X = \frac{Q X}{V} \qquad (6.15)$$

Thus:

$$\mu = \frac{Q}{V} \qquad (6.16)$$

The term Q/V is known as the dilution rate, with units of t^{-1} and it is a measure of the fraction of the culture volume which changes in unit time. Thus a dilution rate of $0.1\,h^{-1}$ means that there is one complete change in culture volume every 10 hours. The reciprocal of the dilution rate is known as the mean cell residence time and represents the average time an organism remains in the reactor. This term is most widely used in wastewater engineering where it is known as sludge age and designated with the symbol θ Equation (6.16) is of profound importance to the understanding of bacterial growth under continuous cultivation as it shows the dependence of specific growth rate on the flow rate of substrate into a reactor. Thus by controlling the flow of medium into a reactor at a constant, known rate, an organism may be grown at any required specific growth rate, provided that this does not exceed its maximum specific growth rate (μ_m). This facility circumvents the problems experienced by Monod in his attempts to correlate the effects of limiting substrate concentration on specific growth rate, as cultures of microorganisms can be established in continuous culture at a number of different dilution rates and the concentration of rate-limiting nutrient determined. The nature of the rate-limiting nutrient may also be varied easily by reducing the concentration of that compound in the medium for which a nutrient limit is desired.

An equation to describe the concentration of the rate-limiting substrate in the effluent may now be obtained by substituting the right-hand side of the Monod equation (6.8) for the term μ in equation (6.16), and by replacing the term Q/V with the reciprocal of the sludge age $(1/\theta)$, thus:

$$\frac{1}{\theta} = \frac{\mu_m S_e}{K_s + S_e} \tag{6.17}$$

this may now be rearranged in terms of S_e:

$$S_e = \frac{K_s}{(\mu_m \theta - 1)} \tag{6.18}$$

In the same way that a mass balance for changes in biomass concentration in the reactor was established in equation (6.12), we can perform a similar mass balance for changes in substrate concentration. As a result of microbial growth in the reactor, the concentration of substrate will decline to satisfy this growth as described by equation (6.6), in addition substrate is also being lost in the overflow from the reactor. However, both of these losses are being offset by the substrate present in the influent and a substrate balance in the reactor may be expressed as

Rate of change of   Rate of inflow   Rate of outflow
substrate concentration = of new substrate − of substrate   (6.19)
in the reactor

+ Rate of substrate
utilisation

Thus:

$$\frac{ds}{dt} V = QS_i - QS_e + \frac{\mu X}{Y} V \tag{6.20}$$

$$= Q(S_i - S_e) \frac{\mu X}{Y} V \tag{6.21}$$

Dividing by V and assuming a steady state exists $(dS/dt = 0)$:

$$\frac{Q}{V}(S_i - S_e) = \frac{\mu X}{Y} \tag{6.22}$$

As the term Q/V has been shown to equal to specific growth rate (6.16), these two terms cancel yielding:

$$X = Y(S_i - S_e) \tag{6.23}$$

The effluent substrate has previously been described by equation (6.18) and if this is now substituted in the above equation:

$$X = Y\left[S_i - \frac{K_s}{(\mu_m \theta - 1)} \right] \tag{6.24}$$

Utilizing equations (6.18) and (6.24), the steady-state effluent

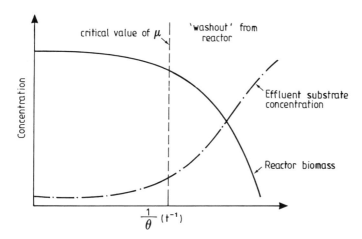

Figure 6.8 The effects of solids retention time on effluent substrate concentration and the concentration of biomass in a continuous-flow stirred tank reactor.

concentrations of biomass and rate-limiting substrate may now be calculated providing that the three basic growth parameters—yield, saturation coefficient and maximum specific growth rate—are known. The effect of the solids retention time on the concentration of organisms in the reactor and the substrate concentration in the effluent, as predicted by equations (6.18) and (6.24), is depicted graphically in Figure 6.8.

Further inspection of equation (6.18) reveals that the effluent substrate is controlled solely by the solids retention time and is independent of the influent substrate concentration. If this is examined in conjunction with equation (6.24) it is apparent that although theory predicts that effluent substrate will not increase in response to increases in effluent substrate concentration, there will be an increase in the solids concentration in the reactor. Common sense suggest that such a situation cannot continue indefinitely, and indeed there is a finite limit as to the amount of solids which may accumulate in the reactor before several other factors, for instance the mass transfer of oxygen between the gaseous and liquid phase and solubility of substrate, become rate limiting. In addition these equations are not meant to imply that a sudden shock loading of substrate will not influence the magnitude of the effluent substrate concentration. Both equations (6.18) and (6.24) were derived on the basis that a steady state was in operation; consequently there will be a lag period before the growth rate accommodates to the new substrate concentration and the cell concentration in the reactor increases. This observation has been confirmed experimentally and it has been shown that the specific growth rate experiences a dynamic lag in responding to changes in the concentration of rate-limiting substrate in the culture vessel. A small increase in growth rate is generally observed immediately, but further increases to a new higher steady-state value may take several solids retention times to accomplish. A lag in the responses of the specific growth rate of the population to rapid increases or decreases in substrate concentration is known as a 'hysteresis effect', whereas the 'inertial phenomena' describes a population in which the cell concentration rises initially to a value higher than that predicted by equation (6.24) before attaining the predicted steady-state concentration. It is possible to derive dynamic models of these responses assuming that the relationship between the variations of substrate with time are known. This allows the integration of equation (6.24) (which has been achieved using a Laplace transform) to yield a dynamic response model. However, the complexity of the stochastic equations necessary to predict the diurnal and seasonal variations observed in substrate concentrations to sewage works has meant that this approach has received little attention.

6.4 MICROBIAL GROWTH AT LONG SOLIDS RETENTION TIMES

The aim of all industrial processes which utilise continuous-culture techniques is to reduce the solids retention to a minimum while maintaining a high level of product yield and quality. This will obviously result in either a reduction in reactor size for a given volume of product produced or an increased yield of product for a given reactor volume. Similarly, the aim of modelling in wastewater treatment is to produce the smallest reactor size compatible with the required degree of treatment. However, as a result of the composition of the influent substrate to wastewater treatment plants, which is very low in carbon source as compared to those employed in industrial processes (the glucose equivalent of domestic wastewater is 300 mg glucose/l as compared with 30 000 mg/l for the substrate in single-cell protein production), solids retention times of from 3 to 30 days are required as compared with typical values of 3–10 hours for industrial processes. Under such conditions, bacterial growth shows many deviations from the models developed previously, and a full understanding of the consequences of growth at long solids retention times is required in order to make appropriate modifications to these models.

Maintenance energy

Implicit in the derivation of the equations for microbial growth was the assumption that the yield coefficient was constant for a given set of environmental conditions. Thus inspection of equation (6.7) which relates yield to the growth rate and specific substrate utilisation rate, reveals that a plot of u vs q will be linear passing through the origin

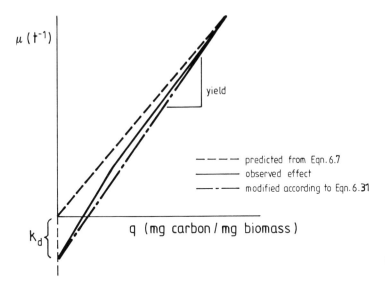

$\mu\ (t^{-1})$

yield

– – – – – predicted from Eqn. 6.7
————— observed effect
—·—·— modified according to Eqn. 6.31

q (mg carbon / mg biomass)

k_d

Figure 6.9 The influence of the solids retention time in a reactor on the value of the yield coefficient.

with gradient Y, hence the yield coefficient is independent of growth rate (Figure 6.9). Although this relationship has been confirmed for organisms grown at high growth rates, at low growth rates the yield is observed to vary markedly, and significantly lower values are measured for this coefficient under such conditions. This dependence of yield on specific growth rate has been taken as an indication of a microorganism's requirements for energy for purposes which are not associated with growth. This energy requirement has been termed 'maintenance energy' and is utilised for a variety of essential cellular processes including the turnover of cellular constituents, the maintenance of a pool of intracellular metabolites against a concentration gradient and the preservation of the correct intracellular ionic composition and pH. The concept of maintenance energy suggests that, during growth, the consumption of energy is used for two processes, cell synthesis and cell maintenance. The energy used for cell synthesis is obviously a growth-related process, as cells growing faster will require more energy; however, the fraction required for maintenance purposes will be growth independent as all viable microorganisms require the correct intracellular environment regardless of their growth rate. This helps us to understand why the deviation in cell yield is more noticeable at low growth rates, as the fraction of energy used for maintenance purposes will be higher. An expression has been derived to describe the above physiological explanation for the deviation observed in Figure 6.9, in which the relationship between the growth rate and the yield value are expressed in terms of the substrate being consumed (q). The equation contains a growth-associated term for substrate consumption (q_g) and a non-growth-associated term for substrate consumption (q_m) and it takes the form:

$$q = q_g + q_m \tag{6.25}$$

If equation (6.25) is now divided by the specific growth rate (μ):

$$\frac{q}{\mu} = \frac{q_g}{\mu} + \frac{q_m}{\mu} \tag{6.26}$$

From equation (6.7):

$$\frac{1}{Y} = \frac{1}{Y_g} + \frac{q_m}{\mu} \tag{6.27}$$

It is clear from equation (6.27) that the yield is now dependent on the growth rate. The equation also helps to explain why different organisms may exhibit widely different growth yields as the growth yield is very dependent on the amount of substrate consumed for maintenance purposes. The value Y_g is often referred to as the true yield (Y_T), as it represents the actual amount of organisms which have been synthesised from a given amount of substrate. Conversely, Y is referred to as the observed yield (Y_{obs}) and it is the yield value determined experimentally. Equations (6.26) and (6.27) both assume that the maintenance energy requirements are constant at all growth rates; however, numerous deviations have been found for certain growth conditions. This has led to the suggestion that the capacity of bacteria to regulate the rate at which ATP is produced by catabolism, and the rate at which it is consumed by anabolism, may be limited. Thus under certain conditions more ATP is produced than is required for anabolism, and this excess production must be burnt up by non-growth-associated processes. This phenomenon is known as uncoupling, as the link between energy production and growth has been 'uncoupled'. The occurrence of uncoupling is reflected in a high maintenance coefficient and corresponding low yield. It has been demonstrated under many different conditions, for instance:

1. In minimal media;

2. With substrates as carbon and energy source which are not energy limiting for growth;

3. In the presence of certain inhibitory compounds;

4. In media containing suboptimal amounts of an essential growth factor.

The ability to uncouple energy production from growth would be extremely useful in many wastewater treatment processes, as the handling and disposal costs of the biomass produced as a result of substrate removal may account for as much as 60% of the total plant operating costs.

Low viability

The equations to describe the growth of microorganisms in continuous culture take no account of the viability of the microbial
population, and indeed such an approach is not necessary because the
equations have been derived to describe the overall behaviour of a
large population at steady state. The implications to growth for a
population which contains a significant fraction of non-viable
organisms is that the rate of growth of the viable fraction of the
population must be greater than the net observed growth rate. The
true growth rate of the viable fraction may be expressed by
incorporation of an index of viability α, which is the fraction of viable
organisms in the population:

$$\mu_v = \frac{\mu}{\alpha} \qquad (6.28)$$

where μ_v is the true population growth rate and μ the observed
growth rate calculated from equation (6.17). Values for α in the
activated sludge process range from 0.01 to 0.1, and this suggests that
many of the organisms in the reactor will be in, or approaching, the
stationary and death phase of the growth curve as illustrated in
Figure 6.1. A feature of this phase of growth is that the organism
undergoes the autodigestion of its own intracellular material cul-
minating in cell lysis. This phenomena is frequently referred to as
endogenous metabolism, although strictly speaking endogenous
metabolism describes a strategy which is adopted by microorganisms
when deprived of an external source of nutrients and a preferred term
is 'organism decay'. Typical endogenous substrates include the
intracellular amino-acid pool and cellular RNA.

In addition a number of microorganisms synthesise storage
products such as poly-2-hydroxy butyrate (phb), glycogen or polyphosphates which are a rich source of reducing equivalents for use in
times of nutrient depletion. Possession of such compounds usually
confers a survival advantage on the organism. Upon the depletion of
the intracellular reserve material the organism loses viability and
undergoes cell lysis, and this results in a significant decline in the cell
mass of the population. Per unit of time, the net loss due to
endogenous respiration is proportional to the biological active mass
and hence the rate of change of live mass due to endogenous
metabolism can be expressed as

$$-\frac{dX}{dt} = k_d X \qquad (6.29)$$

where k_d is known as the decay coefficient with units of mass cells
degraded/mass cells present $\times t$. Thus when microorganisms are
grown under conditions of low viability (long solids retention times
and low substrates utilisation rates), equations to describe their

growth must be modified to take account of both maintenance energy requirements and organism decay. These terms are considered proportional to the mass of organisms and therefore equation (6.6) is written as

$$\frac{dX}{dt} = Y\frac{dS}{dt} - k_d X \tag{6.30}$$

and equation (6.7) becomes

$$\mu = Y_{obs} q - k_d \tag{6.31}$$

The effect of this modification to equation (6.7) is illustrated in Figure 6.9.

The relationship between maintenance energy and the decay coefficient can be described by the equation:

$$k_d = Y_T m \tag{6.32}$$

where m is the maintenance coefficient (g cells/g COD $\times t$) and Y_T the true yield coefficient (g cells/g COD).

Having arrived at modified equations to describe growth at long solids retention times, it is now necessary to reconstruct the biomass and substrate mass balance equations (6.18 and 6.24) incorporating these modified equations. Thus the new equation for effluent substrate concentration becomes

$$S_e = \frac{K_s(1 + k_d\theta)}{\mu_m\theta - (1 + k_d\theta)} \tag{6.33}$$

and effluent biomass concentration is described by

$$X = \frac{Y_{obs}(S_i - S_e)}{(1 + k_d\theta)} \tag{6.34}$$

Obviously as k_d approaches zero as a limit, equations (6.33) and (6.34) reduce to equations (6.18) and (6.24).

6.5 PREY–PREDATOR RELATIONSHIPS

The equations above describe both the growth of pure cultures of microorganisms and the outcome of competition between two organisms for a growth-limiting substrate. Mixed culture phenomena, however, are not merely composites of the pure culture behaviour of the organisms present; the performance of a complex microbial process (such as wastewater treatment) depends upon the interactions between its component species. The establishment and interpretation of results of complex mixed-culture systems presents a formidable research problem because of the large number of variable parameters involved. However, this approach is now starting to gain impetus as a tool to aid in the development of better models to describe various aspects of wastewater treatment processes. This section will cover briefly one aspect of mixed-culture modelling which, in the view of the author, will become increasingly important in wastewater treatment, that is, biological phenomena which may be described by prey–predator relationships. Examples of such phenomena might include pathogen removal by protozoa and bacteriophage in waste stabilisation ponds and the removal of free-swimming bacteria in the activated sludge process by ciliated protozoa. In addition, the effect of predation in any wastewater treatment process is to cause a reduction in the biomass concentration which will result in a reduction in the observed yield. As witnessed in the previous section, much effort has been expended in deriving equations to account for these discrepancies in yield and this has culminated in the establishment of the decay coefficient. In view of the large number of predators observed in wastewater treatment processes, it seems probable that they influence the magnitude of the decay coefficient to a far greater extent than does organism decay.

Equations to describe the growth relationships between predator and prey, generally comprise two terms, one of which describes the growth of the predator population and one which describes the growth of the prey. Thus for the predator:

Rate of change = Rate of predator   − Rate of loss from
of predator   biomass production   culture vessel
population   (growth)   (wash-out)

$$(6.35)$$

The growth of predator is described by a Monod equation which assumes that the prey is the rate-limiting substrate:

$$\frac{dP}{dt} = \frac{\mu_{mp} + H}{K_{sp} + H} P - \frac{P}{\theta} \qquad (6.36)$$

where μ_{mp} is the maximum specific growth rate of the predator in the presence of saturating prey and K_{sp} the predator's saturation constant for the prey. A similar equation can be developed for the change in prey population:

Rate of change   = Rate of prey   −Rate of culture
of prey population   biomass production   loss (wash-out)
    (growth)
    − Rate of prey biomass utilisation
    by the predator (predation)

substituting:

$$\frac{dH}{dt} = \frac{\mu_{mH} + S}{K_{SH} + S} H - \frac{H}{\theta} - BHP \qquad (6.38)$$

where μ_{mH} is the maximum specific growth rate of the prey, K_{SH} the saturation constant of the prey and S the growth-limiting substrate.

The term **BHP** which describes the rate of growth-limiting substrate utilisation is equivalent to the term uX/Y and can be replaced by

$$\frac{\mu_{mP} + H}{(K_{SP} + H)} \frac{P}{Y} = \text{BHP} \tag{6.39}$$

where Y represents the growth yield of the predator on the prey with units g predator biomass produced/g prey biomass utilised. BHP can now be replaced in equation (6.38):

$$\frac{dH}{dt} = \frac{\mu_{mH} + S}{K_{SH} + S} H - \frac{H}{\theta} - \frac{\mu_{mP} + H}{K_{SP} + H} \frac{P}{Y} \tag{6.40}$$

Equations (6.36) and (6.40) can now be used to model a typical prey–predator relationship. Figure 6.10 shows the predicted results of

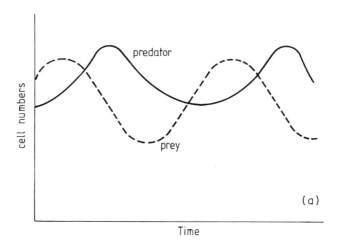

(a)

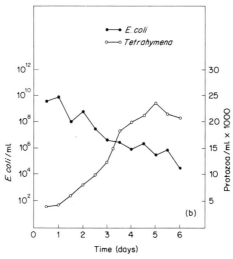

(b)

Figure 6.10 Oscillations of prey and predator populations: (a) as predicted by equations (6.38) and (6.40). (From Lynch and Poole, 1979. Reproduced by permission of Blackwell Scientific Publications Ltd.); (b) as observed for the predatory protozoa *Tetrahymena pyriformis* isolated from activated sludge, preying on the bacterium *E. coli.*

this model for a hypothetical prey–predator pair and oscillating population sizes are observed, with the predator population lagging one-quarter out of phase behind the prey population. A typical laboratory experiment of a predator/prey population isolated from an activated sludge plant is also illustrated to show the validity of the concepts described in this model. Determination of the appropriate coefficients for the prey and predator would allow the model to be calibrated and thus better agreement would be obtained.

7 Modelling and Design of Biological Reactors

7.1 INTRODUCTION

Mathematical models of wastewater treatment processes can be of great benefit to wastewater engineers in several ways. By allowing the testing of a number of potential treatment options, configurations and operational strategies, they are an invaluable aid in selection of the most appropriate treatment system for a given situation. In addition, for existing treatment plants, models can be used to investigate the effects of alternative operational strategies, in order to improve treatment efficiency and reduce operating costs.

The derivation of mathematical models which can accurately describe and predict the events occurring during a wastewater treatment process, requires the selection of suitable equations to describe both the hydraulic regime within the reactor and also the growth of microorganisms on organic and inorganic substrates. Many of the equations which will be utilised in model development have been explained and derived in earlier chapters, and the overriding aim should always be to obtain the simplest model possible which is capable of performing the functions required of it. A number of preliminary selections have to be made before the process of model building can begin and the most important of these are given under the headings below.

(a) Hydraulic flow pattern

The hydraulic flow pattern can vary between extremes of completely mixed and plug flow. For existing reactors the choice of hydraulic flow pattern is determined by tracer studies which reveal the actual dispersion characteristics of the reactor under consideration. For reactors which have yet to be constructed, a choice has to be made between the two extremes. The advantages and disadvantages of the two flow patterns must be considered with a view to the type of wastewater the reactor will treat. For completely mixed reactors the effluent substrate concentration is the same as that in the reactor. This means that influent wastewater is rapidly dispersed throughout the reactor and its concentration is reduced. This feature proves advantageous at sites where periodic discharges of concentrated industrial wastes are received. The rapid dilution of the waste means that the concentration of any toxic compounds present in it will be reduced, and thus the microorganisms within the reactor may not be affected by the toxicant. Therefore, completely mixed reactors generally produce an effluent of more uniform quality which does not undergo large fluctuations in response to fluctuation in influent strength. This property of uniformity also applies to the rate of oxygen uptake by the microorganisms in the reactor. As this is reasonably consistent throughout the reactor, the sizing of aerators and selection of operating regime to meet oxygen requirements are much simplified.

Plug-flow reactors on the other hand have a decreasing substrate concentration gradient from inlet to outlet, which means that toxic compounds in the influent remain undiluted during their passage along the reactor, and this may inhibit or kill many of the microorganisms within the reactor. The substrate concentration gradient along the reactor also means that the oxygen demand along the reactor will vary. This makes the selection and operation of aerators more difficult. On the other hand, the increased substrate concentration means that rates of reaction are increased, and for two reactors of identical volume and hydraulic retention time, with the same BOD removal constant, a plug-flow reactor will show a greater degree of BOD removal than a completely mixed reactor.

A further advantage of an increased substrate concentration is that it appears to select for floc-forming microorganisms at the expense of the filamentous organisms associated with activated sludge bulking. Because of the importance of sludge settlement in ensuring effective and efficient reactor performance, this last factor has tended to swing the balance in favour of plug-flow reactors. The relationship between the dispersion number, tank geometry and fluid velocity is given by the equation:

$$\delta = \frac{D \cdot W \cdot H}{L \cdot Q \cdot (1 + R)} \tag{7.1}$$

where D is the dispersion coefficient (m^2/s), W the reactor width (m), H the reactor depth (m), L the reactor length (m), Q the average flow rate (m^3/s) and R the recycle ratio.

The dispersion coefficient measures turbulence in the aeration tank and is thus a function of the aeration system. It has been calculated at $0.068\ m^2/s$ for diffused air systems. To ensure good sludge settlement, reactors should be designed with dispersion numbers < 0.06 (see Chapter 9). In order to achieve this the following reactor parameters are recommended:

$$2 < W < 20(m)$$

$$2.4 < H < 6.0(m)$$

$$28 < L < 500 \text{(m)}$$

$$0.7 < R < 1.5$$

The dispersion number is related to the number of tanks in series by combining equations (3.25) and (3.26), thus:

$$\delta = \frac{1}{2N} \tag{7.2}$$

This is combined with equation (7.1) to yield:

$$N = \frac{7.4LQ(1 + R)}{WH} \tag{7.3}$$

This is a convenient way to represent an existing activated sludge reactor, as it allows appropriate mass-balance equations to be set up based on the flow diagram, shown in Figure 3.5, for any reactor configuration. This also allows many other process configurations such as step-feeding and contact stabilisation to be modelled.

(b) Substrate removal models

The simplest technique for modelling substrate removal in waste-water reactors is to observe the rate at which substrate is removed with time and to define this by means of a simple rate-removal constant (zero, first or second order). More satisfactory, however, is to adopt a mechanistic approach. This typically utilises the Monod model for microbial growth on rate-limiting substrates. This may then be expanded to yield information on the amount of surplus biomass production and the reactor oxygen requirements, together with important reactor operating parameters.

In certain cases, however, the Monod approach proves inadequate and more sophisticated models must be employed, which consider the biodegradability of the substrate and also the metabolism of the sludge microorganisms.

(c) Process configuration

A typical wastewater treatment plant achieves an effluent BOD of 20 mg O_2/l, and the mixed population of microorganisms in the reactor has kinetic coefficients of 2.5 d^{-1} for μ_m and 100 mg/l for K_s. Thus from equation (6.7) the required reactor volume to give the necessary retention time is

$$\frac{Q}{V} = 2.5 \times \frac{20}{100 + 20} \qquad V = \frac{Q}{0.42} \tag{7.4}$$

Consequently, in order to permit the microorganisms adequate time to grow and divide, a reactor volume greater than twice the daily flow would be required. Obviously it is not practical to provide such large reactors, but at the same time, if the microorganisms do not remain in the system for the amount of time predicted by equation (6.7), then they will not undergo division and will thus wash out of the system (Figure 6.8). In order to circumvent this problem, wastewater treatment systems divorce the retention time of microorganisms in the reactor from the retention time of the fluid in the system. This permits large solids retention times at short hydraulic retention times, with a consequent reduction in reactor volume. This is achieved in one of two ways, either by fixing or immobilising the microorganisms on to a solid support and passing the liquid phase through the solids phase, or alternatively, by separating the solid phase from the liquid phase, recycling the solids phase back to the reactor vessel and discharging the liquid phase. The first system involving immobilised biomass is known as a fixed-film reactor and is exploited in trickling filters and rotating biological contactors. The second technique utilises a sedimentation tank to remove the biomass phase from the liquid phase before recycling a fraction of the solids back to the main reactor. Solids recycle is the distinguishing feature of the activated sludge process and its many modifications. Since the solid phase remains in suspension in the liquid phase and is not immobilised, this is also known as a suspended growth process.

(d) Selection of flow and load values

Wastewater treatment systems receive influents which are highly variable in terms of both flow rate and influent composition. These variations are both diurnal and seasonal and may vary by a factor of three or more around a mean value. Typical daily variations in influent flow rate and strength are illustrated in Figure 7.1.

It would seem logical, therefore, to select an unsteady-state model to describe reactor behaviour. However, this is rarely the case in practice since unsteady-state models frequently have no analytical solution and their complexity leads to a lack of applicability. It is more usual to adopt a steady-state approach to modelling, in which discrepancies resulting from variations in influent are accounted for by means of safety factors. The major problem with such an approach is in predicting the flow and load changes which are likely to occur during the life of the plant. This is mainly due to the lack of historic and current flow data. In order to forecast the flow to a sewage works, a number of parameters which make up the flow are considered, namely: per capita domestic water and unmeasured commercial water returned to the sewer ($[PC(D + UMC)_R]$); trade flow to sewer (E) and infiltration (I). Flow is normally described in terms of the dry

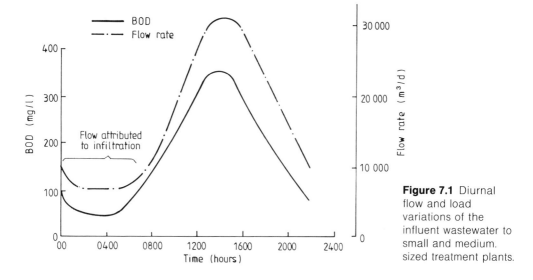

Figure 7.1 Diurnal flow and load variations of the influent wastewater to small and medium. sized treatment plants.

weather flow (DWF), and thus

$$DWF = \text{population} \times PC(D + UMC)_R + E + I \qquad (7.5)$$

Trends in population are generally derived from the results of censuses and surveys; projections over a 25-year period can often be made from these, with some reliability.

For an existing treatment plant, the DWF has been defined as 'the average daily flow during seven consecutive days without rain (excluding a period which includes public or local holidays) following seven days during which the rainfall did not exceed 0.25 mm on any one day'. However, in many areas the rainfall pattern means that it is not possible to measure a DWF according to the above definition. A simpler method is to measure the daily flow for all the days during the year when no rain fell. The DWF is then given by the median value of these flow rates. A figure of 3 × DWF is widely used when designing a new works, and this is termed the flow to full treatment (FTFT), thus:

$$FTFT\ (m^3/d) = 3PG + 3E + I \qquad (7.6)$$

where P is the domestic population, G the domestic water usage $(m^3/\text{person d})$, E the industrial flow (m^3/d) and I the infiltration (m^3/d).

Infiltration represents entry of groundwater into the sewers and many rule-of-thumb techniques are available for its determination. Typical values are in the range 20–50 l/km d for each millimetre diameter of sewer. It is also often approximated as the flow to the works recorded between the hours of 2 and 4 a.m. assuming that the contribution from industrial activity is small.

In addition to flow variations, the organic load also varies widely.

The total load to a works is given by

$$\text{Load in BOD (kg/d)} = \text{Population} \times \text{Domestic load per capita}$$
$$+ \text{Industrial load} \qquad (7.7)$$

The domestic load per capita falls within the range 23–78 g BOD_5/d and varies both within a country and between countries (Table 7.1). It is unlikely to change significantly over a 20-year period. The industrial load, however, can show large changes, as the industrial patterns within an area changes. Typical flow and loads for a range of sewage discharges are given in Table 7.2 and this illustrates the problem of forecasting load variations.

The following sections describe some of the more influential models which have been developed to describe a range of treatment options.

Country or region	BOD_5 per capita daily in sewage (g)
Brazil (São Paulo)	50
France (rural)	24–34
India	30–55
Kenya	23–40
Nigeria	54
Southeast Asia	43
UK	50–59
USA	45–78
Zambia	36

Table 7.1 BOD_5 contributions per capita in urban sewage. (From Feachem et al, 1983. Reproduced by permission of the World Bank.)

Source	Volume of sewage (l/head of population day)	Population equivalent
Small domestic housing	120	1.0
Luxury domestic housing	200	1.7
Hotels with private baths	150	1.3
Restaurants (toilet and kitchen wastes per customer)	30–40	0.3
Camping sites with central bathrooms	80–120	0.7–1.0
Camping sites with limited sanitary facilities	50–80	0.4–0.7
Day-schools with meals service	50–60	0.4–0.5
Boarding-schools, term time	150–200	1.3–1.7
Offices, daywork	40–50	0.3–0.4
Factories—per 8 h shift	40–80	0.3–0.7

Table 7.2 Guidelines for sewage discharge from various types of accommodation.

7.2 SUSPENDED GROWTH SYSTEMS

Completely mixed reactors with recycle

(a) Sludge age

Equations for a completely mixed reactor with no recycle were derived in Chapter 6, and for the solids retention time routinely employed in wastewater treatment plants, equations (6.33) and (6.34) were obtained. The same procedure is followed when recycle is incorporated, with slight modifications to the initial mass-balance equations. Thus from the schematic diagram in Figure 7.2 a biomass mass balance around the reactor and clarifier is expressed as

Rate of change of = Rate of inflow + Rate of biomass
biomass in reactor   of biomass   production

$\qquad$ − Rate of biomass
$\qquad\qquad$ removed.$\hspace{3cm}$(7.8)

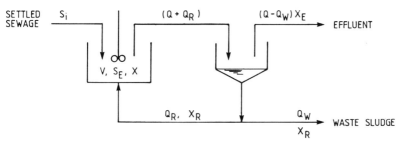

Figure 7.2 Schematic diagram for a completely mixed activated sludge reactor, incorporating sludge recycle from a secondary sedimentation tank.

For a reactor of volume V, and assuming negligible influent solids,

$$\frac{\mathrm{d}x}{\mathrm{d}t}V = \mu X V - [(Q - Q_w)X_e + Q_w X_r]\hspace{2cm}(7.9)$$

dividing by V,

$$\frac{\mathrm{d}x}{\mathrm{d}t} = \mu X - \frac{[(Q - Q_w)X_e + Q_w X_r]}{V}\hspace{2cm}(7.10)$$

at steady state $\mathrm{d}x/\mathrm{d}t = 0$. Therefore,

$$\mu = \frac{(Q - Q_w)X_e + Q_w X_r}{V X}\hspace{2cm}(7.11)$$

substituting for sludge age:

$$\theta = \frac{V X}{(Q - Q_w)X_e + Q_w X_r}\hspace{2cm}(7.12)$$

It is apparent upon inspection of equation (7.12) that the sludge age is the ratio of biomass in the reactor to the rate at which biomass is lost from the clarifier (both in the effluent and by wasting). The term 'sludge age' is often used interchangeably with the term 'mean solids retention time' (θ_m); however, strictly speaking θ_m should contain a decay term to account for the solids which are lost due to auto-digestion and lysis, thus

$$\theta_m = \frac{VX}{(Q - Q_w)X_e + Q_w X_r + k_d XV} \tag{7.13}$$

(b) Food:microorganism ratio

If a substrate mass balance around the reactor is now performed then:

Rate of change   = Rate of substrate   − Rate of substrate
of substrate   inflow   outflow
                                                          + Rate of substrate
                                                             removal   (7.14)

$$\frac{ds}{dt}V = QS_0 + Q_r S_e - \left[\frac{\mu XV}{Y} + (Q + Q_r)S_e \right] \tag{7.15}$$

Thus

$$\frac{ds}{dt} = \frac{Q}{V}(S_0 - S_e) - \frac{\mu X}{Y} \tag{7.16}$$

at steady state when $ds/dt = 0$

$$\frac{\mu}{Y} = \frac{Q(S_0 - S_e)}{VX} \tag{7.17}$$

from equation (6.7), μ/Y may be replaced by q and thus

$$q = \frac{Q(S_0 - S_e)}{VX} \tag{7.18}$$

The term q represents the amount of substrate that a given amount of biomass is utilising and known as the food:microorganism (F/M) ratio. It has a similar value to the organic loading rate and differs only in that the latter term is always slightly greater, as it does not account for BOD which leaves in the effluent.

(c) Reactor solids concentration

The F/M ratio is related to the sludge age by equation (6.31) and if q is replaced by the term $(1/\theta + k_d)/Y$ in equation (7.18) then:

$$\frac{1/\theta + k_d}{Y} = \frac{Q(S_0 - S_e)}{VX} \tag{7.19}$$

Upon rearranging, an expression for the concentration of solids in the reactor can be obtained:

$$X = \frac{YQ(S_0 - S_e)}{V(1/\theta + k_d)} \qquad (7.20)$$

This equation is often written as

$$X = \frac{Y(S_0 - S_e)\theta}{(1 + k_d)} \frac{}{t} \qquad (7.21)$$

and this serves to illustrate that when the sludge age has the same value as the hydraulic retention time (i.e. there is no solids recycle), equation (7.21) reduces to equation (6.34).

(d) Effluent substrate concentration

It was shown in Chapter 6 that the effluent substrate concentration from a continuous-flow reactor is determined by the solids retention time in the reactor (equation 6.8). This equation is not affected by recycle, except in so far as the long sludge ages employed in wastewater treatment require incorporation of a decay term. This was shown in equation (6.34) which is therefore applicable both in the presence and absence of cell recycle. This equation is particularly important as it is used to calculate the sludge age required for a reactor to achieve a given effluent standard. Inspection shows that this equation is heavily dependent upon kinetic coefficients which effectively define the nature of the wastewater. If it is to be used successfully, pilot-plant studies are required to determine these coefficients accurately.

(e) Reactor volume

Assuming that the appropriate kinetic coefficients (μ_m, K_s, Y and k_d) are known, and that the required effluent substrate concentration has been specified, the sludge age can be calculated from equation (6.34). The reactor volume can now be calculated from equation (7.21) once a reactor solids concentration has been selected by the designer. It is at this stage in the process that the design and performance of the sedimentation tanks must be considered since an increase in the concentration of reactor solids will increase the applied solids loading to the sedimentation tank as governed by equation (3.62). This in turn requires a larger area of sedimentation tank for thickening.

The factors which affect the choice of reactor solids are the efficiency of the secondary sedimentation tank, together with the settling properties of the activated sludge. The clarification function of the sedimentation tank will affect the amount of solids lost in the effluent and is difficult to control, whereas the thickening function will determine the concentration of solids in the recycle sludge (X_R). The importance of the latter is shown from a mass balance around the secondary sedimentation tank:

| Solids entering tank | = Solids leaving in effluent + Solids leaving in recycle | + Solids wasted |

$$(Q + Q_r)X = (Q - Q_w)X_e + Q_w X_r + Q_r X_r \qquad (7.23)$$

The term $(Q - Q_w)X_e + Q_w X_r$ is the denominator in equation (7.12) and can therefore be replaced with the term VX/θ:

$$(Q + Q_r)X = \frac{VX}{\theta} + Q_r X_r \qquad (7.24)$$

dividing by QX:

$$1 + \frac{Q_r}{Q} = \frac{V}{Q\theta} + \frac{Q_r}{Q}\frac{X_r}{X} \qquad (7.25)$$

The term Q_r/Q is the flow rate of the recycled sludge as a fraction of the influent flow rate and is known as the recycle ratio (R), thus:

$$\frac{1}{\theta} = \frac{Q}{V}\left(1 + R - R\frac{X_r}{X}\right) \qquad (7.26)$$

The term X_r/X represents the thickening ability of the secondary sedimentation tank and equation (7.26) shows clearly the interrelationship between the two unit operations of aeration and sedimentation, and how these are affected by the control parameters of sludge age and recycle ratio. This relationship is illustrated in Figure 7.3 for two sludges with different settling properties at a fixed sludge age.

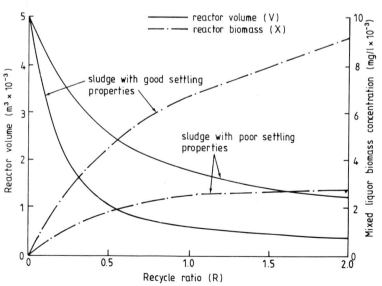

Figure 7.3 The dependence of the reactor volume of an activated sludge reactor, upon the sludge-settling properties and the operating reactor solids concentration, for a range of recycle ratios.

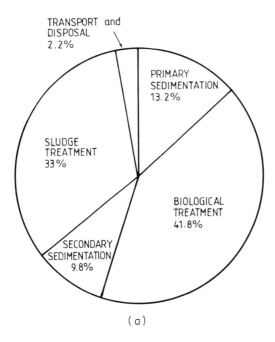

(a)

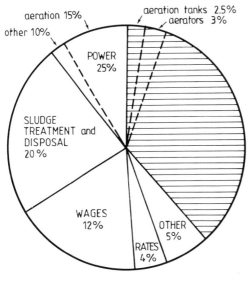

(b)

Figure 7.4 (a) Construction costs associated with the unit operations of a wastewater treatment plant; (b) the annual operating costs of a typical activated sludge wastewater treatment plant.

In addition to its effect on the size of the secondary sedimentation tank, changing the reactor solids concentration will also change the amount of oxygen which is required to satisfy the BOD demand of the sludge. This is discussed in more detail in the section on aeration requirements. It can be seen therefore that increasing the mixed liquor solids concentration in the reactor will have the following effects:

1. The volume required for the aeration basin will be reduced, thus producing savings in construction costs.

2. An increased area of sedimentation tank is required which will incur additional construction costs.

3. For a given combination of reactor volume and mixed liquor solids (Figure 7.3), a decrease in recycle ratio will reduce operating costs as a result of savings in pumping cost.

4. Increasing the mixed liquor solids concentration will cause a reduction in oxygen to satisfy the BOD demand; however, this will be partially offset by the increased energy requirements necessary for mixing.

It is obvious therefore that a large number of combinations of reactor volume, sedimentation tank area and operating conditions can be selected for a given sludge age and effluent BOD. The most suitable combination will be very site specific and depend upon such things as land costs, power costs and labour costs. A breakdown of the annual operating costs for a typical activated sludge plant is illustrated in Figure 7.4. It is at this stage, therefore, that an optimisation is required which can compare the capital and running costs for each of the different combinations, and the optimum combination selected. Standard charts such as 'Cost information for water supply and sewage disposal', prepared by the Water Research

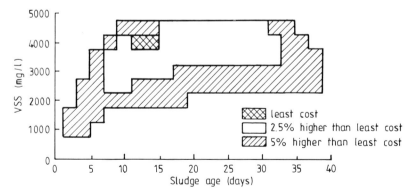

Figure 7.5 Total costs for activated sludge reactors for a range of design mixed liquor solids concentrations and sludge ages. (From Denn, 1984. © 1984 Morton M. Denn.)

Centre in the UK, are available to make the necessary economic comparisons, these are updated regularly to take into account increasing costs.

A typical example of a cost analysis of the various choices of sludge age and microbial biomass is shown in Figure 7.5. This illustrates the flexibility designers have in choosing a range of different values without adversely affecting the final cost.

Plug-flow reactors with recycle

(a) Effluent substrate

As a plug-flow reactor has no longitudinal mixing, the reactions occurring in each element of liquid within the reactor are unique. The concentration of substrate will decrease along the length of the reactor and there will be a corresponding increase in biomass concentration. Derivation of mathematical models for plug-flow reactors is difficult, and simplifying assumptions are required. Thus for the reactor depicted in Figure 7.6 a substrate mass balance around the fluid element of volume dV is written as

$$
\begin{array}{ll}
\text{Rate of change} & = \text{Rate of substrate} \quad + \text{Rate of substrate} \\
\text{of substrate} & \quad \text{inflow} \qquad\qquad\qquad \text{utilisation} \\
& \quad - \text{Rate of substrate} \\
& \quad\;\; \text{outflow} \hspace{4cm} (7.27)
\end{array}
$$

Thus at steady state:

$$0 = (Q + Q_r)S_0 + \frac{\mu X}{Y}\,dV - (Q + Q_r)(S + dS) \qquad (7.28)$$

where S_0 is the substrate concentration to the reactor after dilution with recycle flow. Replacing the term Q_r/Q with the recycle rate, R, and cancelling terms:

$$0 = Q(1 + R)\,dS + \frac{\mu X}{Y}\,dV \qquad (7.29)$$

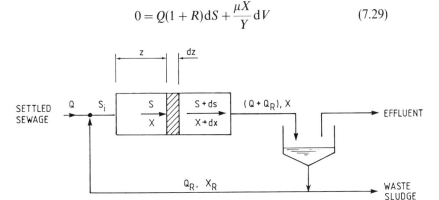

Figure 7.6 Schematic diagram for a plug-flow activated sludge reactor, incorporating sludge recycle from a secondary sedimentation tank.

In order to integrate this equation directly, it is necessary to assume that the biomass concentration within the reactor remains constant. This is a reasonable assumption, as the amount of biomass produced during substrate metabolism along the reactor is small compared to the amount of solids entering the reactor through recycle. If the average biomass concentration along the reactor is denoted as $\bar{X}$, then this replaces X in equation (7.29). If μ is now replaced with the right-hand side of equation (6.8) then:

$$\theta = Q(1 + R)\,dS + \frac{\mu_m S \bar{X}}{Y(k_s + S)}\,dV \tag{7.30}$$

This can now be integrated with $\bar{X}$ as a constant to give

$$(S_0 - S_e) + k_s \ln\frac{S_0}{S_e} = \frac{\mu_m \bar{X} V}{Y(1 + R)Q} \tag{7.31}$$

(b) Reactor solids

A biomass mass balance can also be constructed in the same way, and this is similar to that obtained in equation (7.15)

$$\bar{X} = \frac{YQ\theta(S_0 - S_e)}{V} \tag{7.32}$$

The equations for plug-flow reactors tend to be complex and their solutions difficult, consequently they are very rarely used for design. The equations derived for completely mixed reactors tend to be more widely used, and when a plug-flow reactor is required it is considered as a number of completely mixed reactors in series.

Determination of kinetic coefficients

Biological design of wastewater treatment systems is heavily dependent upon availability of accurate kinetic coefficients. These describe the growth of a mixed population of organisms on the wastewater to be treated and should always be obtained from pilot-plant studies. It is quite simple to obtain these coefficients using a bench-scale activated sludge reactor of the type shown in Figure 7.7. This is operated at a range of different sludge ages and the effluent substrate concentration and mixed liquor suspended solids concentration is measured. With this information the linearised form of equation (7.19), that is,

$$\frac{S_0 - S_e}{Xt} = \frac{1}{Y} \cdot \frac{1}{\theta} + \frac{k_d}{Y} \tag{7.33}$$

may be plotted as $S_0 - S/Xt$ vs $1/\theta$ to yield a straight line of gradient $1/Y$ and intercept k_d/Y.

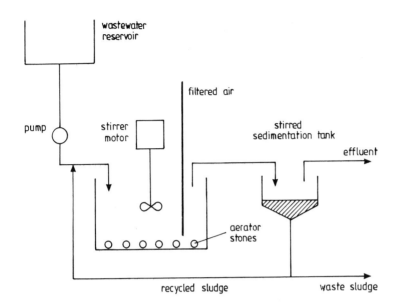

Figure 7.7 A simple bench-scale activated sludge reactor which can be used to determine the kinetic coefficients of a new wastewater.

The coefficients K_s and μ_m are obtained in a similar way from equation (6.7). Here q is replaced with the right-hand side of equation (7.16) and μ is replaced by the Monod equation (6.8) to give

$$\frac{\mu_m S_e}{K_s + S_e} = Y\frac{S_0 - S_e}{tX} \qquad (7.34)$$

this may be linearised as

$$\frac{Xt}{S_0 - S_e} = \frac{K_s Y}{\mu_m}\cdot\frac{1}{S_e} + \frac{Y}{\mu_m} \qquad (7.35)$$

thus a plot of $Xt/S_0 - S_e$ vs $1/S$ is linear with gradient YK_s/μ_m and intercept Y/μ_m.

Typical kinetic coefficients for a range of wastewater are given in Table 7.3.

| | Wastewater | | |
Kinetic coefficient	Domestic	Phenolic	Brewery
$\mu_m(d^{-1})$	2.4–7.2	11	14
$K_s(mg/l)$	50–120	2	30
Y(mg MLSS/mg BOD)	0.5–0.6	0.75	0.6
$k_d(d^{-1})$	0.03–0.06	0.05	0.04

Table 7.3 Typical kinetic coefficients for a range of wastewater.

Aeration requirements

In order to ensure that microorganisms in an activated sludge reactor carry out aerobic metabolism of the substrate to produce energy for growth and division and thus prevent the production of noxious odours, it is important that their aeration requirements are satisfied. However, as aeration is the major energy-using process in activated sludge treatment (up to 80% of the energy budget), it is important to select both an energy-efficient system and also one which matches the aeration requirements of the plant. Oxygen is required within the activated sludge reactor solely for aerobic energy metabolism, in which it acts as a terminal electron acceptor in the oxidation of organic and inorganic material (see Chapter 5). However, metabolism is not the only means by which BOD is removed from the wastewater and it may also be adsorbed on to the sludge flocs, and thus removed with the waste sludge, or it may be incorporated into the sludge microorganisms as new cell material during growth. The oxygen requirement or specific oxygen demand (OV_R) thus has two components: in the absence of an external energy source oxygen is required to carry out essential metabolic processes, or endogenous metabolism, this is known as the endogenous oxygen requirements (OV_E) and is a function solely of the mass of microorganisms in the reactor. If a source of biodegradable substrate is now added, then a further supply of oxygen is required to act as the terminal electron acceptor in microbial respiration (OV_s); this is obviously a function of the amount of BOD removed by oxidation. The oxygen demand has been described empirically by the equation:

$$OV_R(\text{kg/d}) = OV_s + OV_e = 0.5\Delta S + 0.1 X_v \qquad (7.36)$$

where ΔS is the amount of substrate removed (kg BOD_5/m^3 d) and X_v the solids content of the aeration tank (kg/m^3).

A more sophisticated approach has calculated the oxygen demand by relating it empirically to the flow rate ($1/S$), BOD removed and the mixed liquor solids concentration (mg/l), where

$$\text{Oxygen demand (kg O}_2/\text{d)} = 0.0864Q\left[0.75(S_0 - S_e) \right.$$
$$\left. + \frac{5.25 \times 10^{-4} XV}{Q} \right] \qquad (7.37)$$

A more fundamental approach considers that oxygen is required solely to satisfy the ultimate BOD demand of a waste, and therefore, expressing this as BOD_5:

$$\text{O}_2 \text{ requirements (kg O}_2/\text{d)} = 1.47[Q(S_0 - S_e)] \qquad (7.38)$$

where 1.47 is a factor to convert BOD_5 to BOD_u. However, a microorganism uses a fraction of the BOD in a wastewater to produce cellular material (protein, nucleic acids and carbohydrate) during growth and reproduction. This fraction can be calculated if it is assumed that a microorganism has the empirical formula $C_5H_7NO_2$. The oxygen requirements for the synthesis of a microorganism are given by

$$C_5H_7NO_2 + 5O_2 \longrightarrow 5CO_2 + 2H_2O + NH_3 \qquad (7.39)$$
$$\quad (113) \qquad (160)$$

From the molecular weight of the reactants it is seen that formation of 113 kg of microorganisms removes the requirements for 160 kg of oxygen, which is 1.42 kg of O_2 for each kg of solids produced. In other words, this amount of BOD is removed from the influent without the requirement of oxygen. Equation (7.38) is therefore modified to account for this:

$$O_2 \text{ requirement (kg } O_2/d) = 1.47[Q(S_0 - S_e)] - 1.42\Delta X \quad (7.40)$$

where ΔX is the daily solids production (kg/d) and may be calculated if the sludge age is known from the denominator of equation (7.11):

$$\Delta X = \frac{VX}{\theta} \qquad (7.41)$$

For a given sludge age, equation (7.41) shows that as the concentration of solids in the reactor increases, ΔX also increases, and thus from equation (7.40) the oxygen requirements will decrease.

Selection of aerators to supply the calculated oxygen requirements are based on the aerator operating efficiency, expressed as kg O_2/kWh. Consequently the installed aerator power is expressed in kilowatt-hours.

7.3 ATTACHED GROWTH PROCESSES

Although attached growth processes, such as trickling filters and rotating biological contactors, prove simpler to operate than suspended growth processes, it is extremely difficult to model them accurately. This is primarily due to the large number of variables known to influence the process including: filter depth; hydraulic flow and organic loading rate; recirculation ratio; type of filter media; mass transfer of organic material and oxygen from the liquid layer to the attached slime layer; and finally the metabolism of the attached slime layer. The earliest models to describe filter performance were either empirical or simple deterministic models. These simple models have become increasingly complex, and more recent models requiring extensive calibration have a large number of specific coefficients which must be determined before they can be applied.

Simple empirical models

The NRC formula

The National Research Council Model (NRC) was developed in the USA in 1946 and arose from analysis of data collected from treatment plants sited at military installations. For a single-stage filter it takes the form:

$$E = S\frac{1}{1 + 0.014[W/(VR_f)^{0.5}]} \tag{7.42}$$

where E is the fraction of influent BOD removed, W the filter loading (kg BOD/d), V the total volume of filter (1000 m^3) and R_f the recycle factor defined as

$$R_f = \frac{1 + R}{[1 + (1 - P)R]^2} \tag{7.43}$$

where R is the recirculation ratio of recirculated flow:influent flow and P the weighting factor (this generally has the value 0.9).

This equation is most easily utilised by means of a design chart of the type illustrated in Figure 7.8. This chart assumes a value for the influent BOD of 100 mg/l and a flow rate of 5000 m^3/d. As the filter volume is directly proportional to the BOD and flow rate, the filter volume taken from the chart is modified by multiplying it by the ratio of the actual BOD and flow rate of the wastewater to their base values of 100 mg/l and 5000 m^3/d.

For a two-stage filter where the second-stage filter receives settled effluent from the first-stage filter, there is a decline in the treatability of the waste applied to the second-stage filter. The magnitude of this decline depends on the degree of removal occurring in the first stage. This effect was modelled by the NRC who incorporated a retardation factor to the organic loading equal to $[1/(1 - E_1)]^2$, where E is the

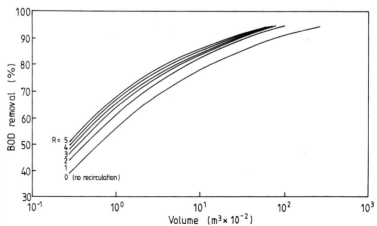

Figure 7.8 Single-stage trickling filter design chart which uses the NRC design equation.

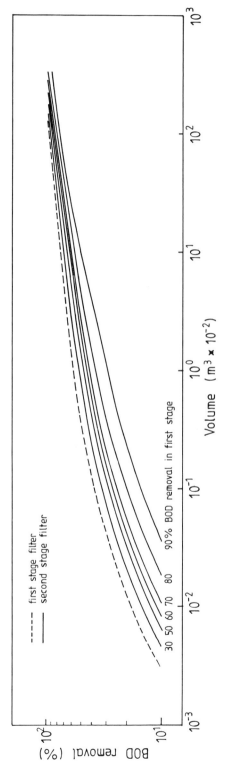

Figure 7.9 Two-stage trickling filter design chart which uses the NRC design equation.

efficiency of the first-stage filter. The complete equation for the second-stage filter thus becomes:

$$E_2 = \frac{1}{1 + 0.014\left(\dfrac{W_2}{VF(1 - E_1)^2}\right)^{0.5}} \tag{7.44}$$

or

$$E_2 = \frac{1}{1 + \dfrac{0.014}{1 - E_1}\dfrac{W^2}{(VF)^{0.5}}} \tag{7.45}$$

Again this equation is applied by means of a design chart (Figure 7.9). The first curve in this chart is used in design of the first-stage filter, whereas the remaining curves are for the second-stage filter and are selected upon the basis of the efficiency achieved in the first stage. It is assumed in this design that a humus tank has been incorporated to receive effluent from the first-stage filter. As with single-stage filters, two-stage filters may also be operated with recirculation. Employment of recirculation will produce a reduction in the total area of filters required. To calculate this reduction, the volume calculated from the design chart in Figure 7.9 for each of the two filters is divided by a recirculation factor. This is found from Figure 7.10, for the recirculation ratio at which each filter will operate. The advantages of a two-stage filter design are that the total volume of the two-stage scheme is always less than that for a single-stage filter, achieving the same effluent quality.

A minimum total volume for a two-stage system is found iteratively by incrementing the first-stage volume, and calculating the total volume required to achieve a given treatment efficiency.

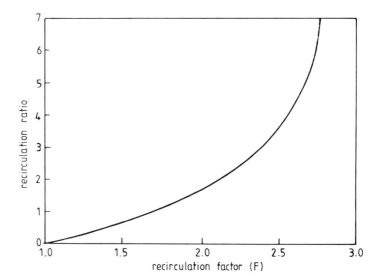

Figure 7.10 Design chart for the calculation of a correction factor to reduce the required volume of filters which operate with recycle.

This second purely empirical method is based on satisfying three criteria, namely that:

1. The BOD load to the filters should not exceed $1.2\,\text{kg BOD/m}^3$ filter d;

2. The hydraulic load, including recirculation, should not exceed $30\,\text{m}^3/\text{m}^2$ filter area d;

3. Recirculation should be such that the BOD entering the filter, including recirculation, is no more than three times the influent BOD. This implies that BOD removal is controlled by recirculation, where removal efficiency is given by

$$E = \frac{R/Q + 1}{R/Q + 1.5} \qquad (7.46)$$

The above three criteria allow calculation of a recirculation ratio from equation (7.46). The filter volume is then determined by dividing the total daily BOD load by $1.2\,\text{kg/m}^3$ d (or less), and the filter area calculated from the hydraulic loading.

The Galler and Gotaas relationship

The approach of Galler and Gotaas in 1964 was to perform multiple regression analysis of data from existing plants, taking L_e as the dependent variable. This resulted in the expression:

$$L_e = \frac{0.464 L_0^{1.19}(1 + R)^{0.28}(Q/A)^{0.13}}{(1 + D)^{0.67}T^{0.15}} \qquad (7.47)$$

This equation predicts that the volume of filter required is a function of flow rate, influent BOD, filter depth, recirculation and the temperature of the wastewater. It may be solved directly for volume in which case it is written in the form:

$$V = \left(\frac{Q^{0.13}L_0^{0.19}[1 + R(1 - E)]^{1.19}}{T^{0.15}(1 + D)^{0.67}(1 - E)(1 + R)^{0.78}}\right) \qquad (7.48)$$

where E is the efficiency of BOD removal.

Generalised filter designs

The Velz equation

One of the earliest models of trickling filter performance was proposed by Velz in 1948, based on the observation that the rate of BOD removal in a filter per interval of depth is proportional to the amount of BOD remaining. This may be expressed in differential form as

$$\frac{dL}{dD} = -kL \tag{7.49}$$

which integrates to

$$\frac{L_D}{L_0} = \exp(-kD) \tag{7.50}$$

where L_0 is the total removable fraction of BOD and L_D the quantity of removable BOD at depth D.

The Velz equation has been modified by numerous workers to take into account the effects of recirculation and temperature. The modified equation has the form:

$$\frac{L}{L_D} = \left\{ (R+1)\exp\left[\frac{k_{20}A_sH\theta^{(T-20)}}{(Q_i(R+1))^n}\right] - R \right\}^{-1} \tag{7.51}$$

where Q_i is the flow before recirculation, divided by filter cross-sectional area $(l/m^2\,s)$, A_s the media specific surface area (m^2/m^3), H the filter height (m), θ the temperature correction factor (generally 1.035) N the empirical flow constant and k_{20} the treatability coefficient $(l/m^2\,s)$.

Because of the complexity of this model, and faced with the problems of a range of model coefficients for the same type of media, it is usually necessary to conduct media and wastewater specific pilot studies, or alternatively select a very conservative filter design.

The Eckenfelder model

As a result of advances in defining trickling filter hydraulics, formulations were developed which related depth, hydraulic loading and media characteristics. Eckenfelder attempted to correlate filter performance to these variables statistically and thus develop generalised design equations. He assumed that BOD removal by trickling filters was proportional to the contact time of the wastewater with the biological slime layer and also to the total active microbial mass in the slime layer. Assuming first-order removal kinetic, this can be expressed as

$$\frac{L_e}{L_0} = \exp(-kXt) \tag{7.52}$$

where k is the a BOD removal rate constant, X the mass of organisms in the slime layer (mg/l) and t the contact time of the wastewater with the slime layer (t^{-1}).

The contact time of the wastewater will be determined by the hydraulic loading rate applied to the filter, the filter depth and the properties of the media within the filter. Residence time can thus be expressed as

$$t = \frac{CA^m D}{Q^n} \qquad (7.53)$$

where D is the filter depth (m) and Q the surface hydraulic loading (m^3/d). Additionally, the surface area of active slime layer per unit volume of filter area will be dependent both upon the type of media and its depth within the filter. If it is assumed, therefore, that the slime layer is proportional to the surface area of the media ($X = k_1 A$) and replacing the expression for contact time (equation 7.53) in equation (7.52), then

$$\frac{L_e}{L_0} = \exp(-k_0 C A^m D/Q^n) \qquad (7.54)$$

in which k_0 is a treatability factor incorporating the surface area of active slime layer/unit volume ($k_0 = k k_1 A$). Thus for a given media with a uniform attachment of slime layer throughout the filter depth and a surface which is constant over time, equation (7.54) simplifies to

$$\frac{L_e}{L_0} = \exp(-k_0 D/Q^n) \qquad (7.55)$$

Although this equation is simpler than the modified Velz equation, it still has a number of coefficients which are media and wastewater specific (k_0 and n). It is necessary to determine these from pilot studies on a rig such as the one illustrated in Figure 7.11. This is filled with appropriate medium and operated at a range of surface loadings with the wastewater which it is intended to treat. Samples are removed at various depths from the filter and the BOD remaining is calculated. If

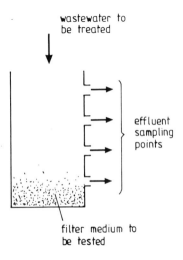

wastewater to
be treated

effluent
sampling
points

filter medium to
be tested

Figure 7.11 Bench-scale trickling filter suitable for determination of the coefficients of a new wastewater.

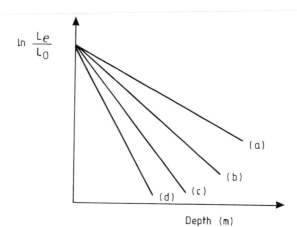

Figure 7.12 Typical data obtained from a bench-scale trickling filter operated at a range of loading rates:
(a) 0.4 kg/m³ d;
(b) 0.2 kg/m³ d;
(c) 0.1 kg/m³ d;
(d) 0.05 kg/m³ d.

equation (7.55) is rearranged in the form:

$$\ln \frac{L_e}{L_0} = -k_0 D Q^{-n} \tag{7.56}$$

then a plot of $\ln L_e/L_0$ vs filter depth will be linear, with a gradient of $-k_0 Q^{-n}$. If a range of loading rates are employed, then a family of curves will be produced (Figure 7.12). Since the gradient (m) of each curve is given by

$$m = -k_0 Q^{-n} \tag{7.57}$$

then

$$\ln m = -\log k_0 + n \log Q \tag{7.58}$$

Thus if the gradient of each of the curves in Figure 7.12 is plotted against the loading rate, then a straight line results with a gradient of n and an intercept of $-k_0$.

It is usual to perform these experiments at a constant temperature, and 20 °C is generally selected. The value of k_0 thus determined is only valid for this temperature and is corrected using an Arrhenius function of the form:

$$k_{0(T)} = k_{0(20)} 1.035^{(-20)} \tag{7.59}$$

7.4 RECENT DEVELOPMENTS IN MODELLING OF BIOLOGICAL TREATMENT SYSTEMS

Modelling of wastewater treatment systems continues to attract a lot of attention and new models are constantly being produced for different aspects of treatment. The widespread availability and usage of microcomputers has proved very useful in this respect and several models are now commercially available which can be used both for treatment plant design and operation. Two of these have proved particularly useful.

(a) Sewage Treatment Optimisation Model (STOM)

This model was developed in the UK between 1970 and 1975 as a collaborative venture between a number of bodies which represent different aspects of the British water industry. Although initially only available on mainframe computers, it is now commercially available for personal computers. The model is semi-empirical and contains a number of modules which relate plant performance and costs to design. The model is very versatile and may be used for such things as: the design of new works and extensions to existing ones; estimating the relationships between treatment costs and the size of a works; and assessing the performance of existing works in order to highlight areas in which a works is under-performing.

Each of the unit processes which comprise a sewage treatment works is represented as a module within STOM and those processes covered are shown in Figure 7.13. Within the modules, mathematical models relate the performance of a process, together with its cost, to the design of that process for defined influent flow rates and strengths. Typical of the many relationships used by STOM is that which

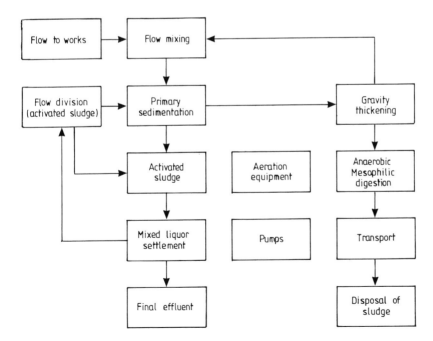

Figure 7.13 The unit processes and auxiliary stages in a wastewater treatment plant, which can be modelled using STOM. (From Spearing, 1987. Reproduced by permission of the Water Research Centre.)

predicts the seasonal average BOD of the clarified effluent from a trickling filter:

$$\left[L_0^{0.5} - \left(\frac{70}{L_0}\right)^{2.2}\right]_{L_e}^{L_0} = \frac{0.4\theta_h C_t C_s}{Q} \qquad (7.60)$$

where L_0 is the seasonal average influent BOD (mg/l), L_e the seasonal average effluent BOD (mg/l), θ_h the treatability coefficient for BOD removal, C_t the temperature correction $(1.07^{(T-15)})$, C_s the seasonal correction to account for cyclical variations in filter biomass and Q the seasonal average BOD of the influent (mg/l).

The mathematical models which are used in the modules are steady-state models and consequently the data which they accept should be monthly average figures. The performances predicted by the model are therefore intended as monthly average performances and STOM does not predict short-term variations caused by fluctuating influents.

Several unique features of the STOM package make it potentially very useful. It may be used in either a fixed mode or an optimising mode. In the former, the performance and costs associated with a

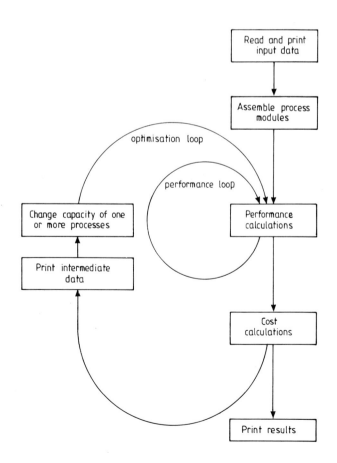

Figure 7.14 Calculation procedure employed by STOM. The inner iterative loop calculates process performance for given values of design variables and outputs the optimum value. (From Spearing, 1987. Reproduced by permission of the Water Research Centre.)

given process configuration are predicted for retention periods which are selected by the operator. In the latter, total treatment costs are minimised by iteratively varying the size of the individual process units to achieve a given effluent standard. The flow diagram used by STOM for these calculation procedures is illustrated in Figure 7.14.

Of course with an ambitious model of this kind there are bound to be limitations, and the usefulness of STOM will depend solely on the validity of the data which it produces. The performance relationships used in STOM were almost exclusively obtained from existing plants within the UK. Although temperature correction facilities are available, there is little evidence to suggest that STOM will prove applicable for use in countries with climates different to that of the UK, without significant modifications.

(b) The International Association of Water Pollution Research and Control Model (IAWPRC)

A major criticism of many of the models used in wastewater treatment is that each model is the result of a single research group and consequently reflects only one view of the process being modelled. In addition, usage of many of the models available is restricted to mainframe computers. In an attempt to rectify this situation, in 1983 the IAWPRC assembled a task group on 'Mathematical modelling for design and operation of biological wastewater treatment'. This group was composed of five representatives with experience in modelling, each from a different country.

The model which they developed was capable of predicting performance of single-sludge activated sludge systems capable of carbonaceous oxidation, nitrification and denitrification. The model recognises eight fundamental processes of importance in single-sludge systems, and for each of these processes a rate equation to represent them is defined (Table 7.4). One important feature of this model is that the concentration of the organic fraction is given in units of COD and that this is subdivided into a number of categories based on its biodegradability. Thus the COD is composed of soluble and particulate fractions and these can be either inert, readily biodegradable and slowly biodegradable. The concentration of the inert soluble fraction is unchanged as it passes through the system, whereas the inert particulate fraction becomes entrapped in the sludge floc and is removed through settlement and wastage. All the readily biodegradable COD is treated as though it were soluble, and this is considered to be the only substrate for the growth of heterotrophic bacteria. The fractions of nitrogen in the influent wastewater are subdivided in similar ways.

The resulting rate equations are fitted into a mass-balance equation which is appropriate for the wastewater treatment system which is being modelled. This includes conventional, step feed and contact

Process	Process rate (M/l^3t)
Aerobic growth of heterotrophs	$\hat{\mu}_H\left(\dfrac{S_s}{K_s+S_s}\right)\left(\dfrac{S_O}{K_{O,H}+S_O}\right)X_{B,H}$
Anoxic growth of heterotrophs	$\hat{\mu}_H\left(\dfrac{S_s}{K_s+S_s}\right)\left(\dfrac{K_{O,H}}{K_{O,H}+S_O}\right)\left(\dfrac{S_{NO}}{K_{NO}+S_{NO}}\right)\eta_g X_{B,H}$
Aerobic growth of autotrophs	$\hat{\mu}_A\left(\dfrac{S_{N,H}}{K_{NH}+S_{NH}}\right)\left(\dfrac{S_O}{K_{O,A}+S_O}\right)X_{B,A}$
'Decay' of heterotrophs	$b_H X_{B,H}$
Decay of autotrophs	$b_A X_{B,A}$
Ammonification of soluble organic nitrogen	$k_a S_{ND} X_{B,H}$
'Hydrolysis' of entrapped organics	$k_h\dfrac{X_s/X_{B,H}}{K_x+(X_s/X_{B,H})}\left[\left(\dfrac{S_O}{K_{O,H}+S_O}\right)\right.$ $\left.+\eta_h\left(\dfrac{K_{O,H}}{K_{O,H}+S_O}\right)\left(\dfrac{S_{NO}}{K_{NO}+S_{NO}}\right)\right]X_{B,H}$
'Hydrolysis' of entrapped organic nitrogen	$\rho_7(X_{ND}/X_s)$

Table 7.4 Process kinetics for carbon oxidation, nitrification and denitrification.

Key to kinetic parameters. Heterotrophic growth and decay: $\hat{\mu}_H$, K_s, $K_{O,H}$, K_{NO}, b_H. Autotrophic growth and decay: $\hat{\mu}_A$, K_{NH}, $K_{O,A}$, b_A. Correction factor for anoxic growth of heterotrophs: η_g. Ammonification: k_a. Hydrolysis: k_h, K_X. Correction factor for anoxic hydrolysis: η_h.

stabilisation activated sludges and can also model such modifications as the Bardenpho process. In order to obtain steady-state solutions to the resulting mass-balance equations, three reasonable assumptions are made:

1. Sludge is wasted from the reactor in proportion to its volume.

2. Solids do not accumulate in the clarifier or leave in the effluent, but are returned to the aerator in the recycle line. Obviously this assumption neglects the effluent solids concentration, but its effect is usually negligible.

3. No reactions occur in the clarifier.

The mass-balance equations which result are non-linear and must be solved iteratively. The initial estimate for the iteration assumes a completely mixed reactor, with first-order removal kinetics. This simplifies the mass-balance equation to a linear form allowing a solution which is the starting-point for the iteration. Using this technique, the effects of diurnal flow variations can be modelled for data inputs of 15-minute intervals.

This model represents a sophisticated approach to the modelling of activated sludge systems performance. Although it will operate on

IBM-PC compatible microcomputers and is reasonably straightfor-ward to operate, it has a major drawback in that a large number of system-specific coefficients are required (Table 7.4) before the model can be used to simulate an actual treatment works. This is an inevitable consequence of attempting a universal model capable of predicting the behaviour of wastewaters which show large variations in composition and are treated under a variety of environmental conditions. It is unlikely that many plants will have the information necessary to use this model, and also unlikely that they will expend the effort required to obtain. The model does, however, specify a number of default values for these parameters which, although they cannot be used for operational purposes, do allow use of the model as a training aid. It is in this mode which it is likely to prove of most value.

8 Nutrient Removal from Wastewaters

8.1 NITROGEN REMOVAL

Introduction

The problems which can arise from the nitrogen pollution of a watercourse, have been considered briefly in Chapter 1. The major problems likely to arise from sewage effluent discharges are nutrient enrichment (eutrophication), with its associated algal blooms and deaeration of the watercourse resulting from oxidation of ammonia to nitrate by the nitrifying bacteria (nitrification). Where nitrification does occur, and the water is abstracted as a potable source, there is the associated problem of nitrate toxicity. Although sewage effluents are not the only source of nitrogen pollution they are a major one, and also the one which is most amenable to control.

The concentrations of nitrogen (and the other essential nutrient, phosphorus) associated with algal blooms are illustrated in Figure 8.1, and consequently the concentration which can be tole-

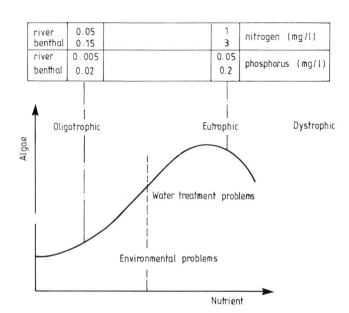

river	0.05		1	nitrogen (mg/l)
benthal	0.15		3	
river	0.005		0.05	phosphorus (mg/l)
benthal	0.02		0.2	

Oligotrophic · · · · · Eutrophic · · · · · Dystrophic

Algae

Water treatment problems

Environmental problems

Nutrient

Figure 8.1 Concentrations of nitrogen and phosphorus in surface water and sediments, associated with environmental problems.

rated in a discharge will depend upon the dilution afforded by the receiving water. Numerous physicochemical methods have been devised for nitrogen removal, such as ammonia stripping, breakpoint chlorination and selective ion-exchange; however, due to their high operating costs and unreliability they have not proved popular. By far the cheapest and most successful technique for nitrogen removal is to exploit the reactions which occur in the biological nitrogen cycle (Figure 8.2) and engineer conditions such that they occur in reactors at wastewater treatment plants.

Settled sewage of domestic origin contains nitrogen either organically bound as protein and nucleic acids, as urea $(OC(NH)_2)$ or as the ammonium ion (NH_4^+). Nitrates and nitrogen are rarely present. The total nitrogen content of a wastewater is often referred to as the total Kjeldahl nitrogen (TKN). This is a reference to the technique by which it is determined, and involves digestion of the sample in strong acid to convert all the bound nitrogen to ammonia. This ammonia is then distilled and collected so that the total ammonia produced can be measured. The nitrogen present in this ammonia represents the organic nitrogen and the free ammonium. If the amount of ammonium present in the original sample is known, then the difference between this and the TKN, given the fraction of organically bound nitrogen.

During any biological treatment process, up to 30% of the total nitrogen is removed in cell synthesis by ammonification, in addition a small fraction of the influent nitrogen will be removed during the sedimentation processes (Table 8.1). In order to remove the remaining ammonia, the nitrogen cycle (Figure 8.2) indicates that this can be achieved by nitrification. In many cases conversion of ammonium to nitrate will provide adequate treatment, with the added advantage that the effluent contains a reservoir of oxygen, in the form of nitrate. However, if eutrophication is likely to be a problem in the receiving

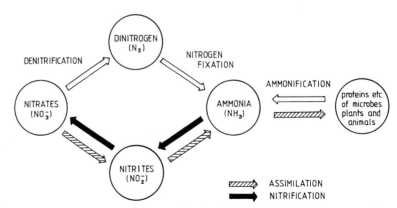

Figure 8.2 Simplified nitrogen cycle showing the nitrogen interconversions which can occur in an aquatic environment.

	Typical concentrations (mg/l)		
Treatment stage	Total nitrogen	Total phosphorous	% Removal
Raw sewage	20–70	6–18	—
Settled sewage	18–60	5–16	5–10
Final effluent (activated sludge)	12–40	4–12	10–30
Final effluent (trickling fillers)	14–45	4–12	8–25

Table 8.1 Typical removal efficiencies of total nitrogen and phosphorous by well-operated, non-nitrifying conventional treatment plants.

water, then it is essential to remove the nitrate as well; the nitrogen cycle shows that this can be removed by encouraging denitrification within the reactor.

3.2 NITRIFICATION

The biochemistry of nitrification has been considered briefly in Chapter 5. The essential feature of nitrification which is of engineering significance is that it is carried out by strictly aerobic, autotrophic bacteria. Two groups of bacteria are required to achieve nitrification: *Nitrosomonas* catalyse oxidation of ammonia to nitrite using molecular oxygen (equation 5.15) whereas *Nitrobacter* further oxidise nitrite to nitrate using oxygen derived from the water molecule (equation 5.16). These reactions provide the nitrifiers with their energy and carbon is assimilated via the Calvin cycle (Figure 5.18) in the form of carbonate or bicarbonate. Assuming a gross cell composition for a typical nitrifying bacteria of $C_5H_7NO_2$, then the overall reaction for the oxidation of ammonia, coupled to the synthesis of new nitrifying bacteria, can be represented as

$$NH_4^+ + 1.83O_2 + 1.98HCO_3^- \rightarrow 0.021C_5H_7NO_2 + 1.041H_2O \\ + 0.98NO_3^- + 1.88H_2CO_3 \quad (8.1)$$

This equation provides three pieces of information of importance in the design of nitrifying wastewater treatment plants.

1. The cell yield (Y) of nitrifiers is very low compared to that of heterotrophs. For every 1 mol of ammonium nitrogen oxidised only 0.021 mol of nitrifier are produced. Converting these into molecular weights, then 18 g of ammonium will yield 113×0.021 g of nitrifier. Thus the yield is $(113 \times 0.021)/18 = 0.13$ g cells/g ammonia oxidised. It is conventional to express these quantities in terms of the nitrogen content of the substrate, thus 1 g ammonium contains $14/18 = 0.77$ g nitrogen. The cell yield is therefore 0.17 g cells produced/g ammonia-N oxidised.

2. There is a large oxygen requirement of 1.83 mol O_2 for every mole of ammonium removed. Thus 1 g of ammonium nitrogen requires 4.2 g oxygen for its removal (note, however, that some of this oxygen will come from the water molecule as shown by equation 5.16).

3. The oxidation of ammonium and removal of bicarbonate during nitrification is associated with a reduction in the alkalinity. Thus each mole of ammonium nitrogen oxidised removes 1.98 mol of bicarbonate. Consequently 8.6 g alkalinity are utilised for each gram of ammonium nitrogen oxidised; if the influent wastewater cannot supply this alkalinity then there will be a drop in the pH of the mixed liquor.

Nitrification by single-stage activated sludge

(a) Calculation of sludge age

As the nitrifying bacteria are energetically inefficient, their growth is very slow and in addition they are sensitive to a wide range of environmental conditions. The most important factors affecting their growth rate (and consequently the sludge age required to ensure that they do not wash out of the reactor), are substrate concentration, temperature, dissolved oxygen and pH. The oxidation of ammonia to nitrite is generally acknowledged to be the rate-limiting step in the nitrification process, and consequently if the growth requirements of this step are met then nitrification will generally proceed to completion in a wastewater reactor, assuming that no inhibitors are present in the settled sewage.

The growth rate of *Nitrosomonas* bacteria has been shown to respond to substrate concentration and dissolved oxygen concentration according to a Monod-type function (equation 6.8) and empirical relationships have been developed to describe the effects of temperature and pH on growth rate. These have been combined into a single equation which can be used to calculate the solids retention time necessary to achieve a given effluent ammonia concentration, under a range of environmental conditions, thus:

$$\frac{1}{\theta} = \mu_{N,max} \frac{(NH_4^+ - N)}{K_N + (NH_4^+ - N)} \frac{DO}{K_{DO} + DO} [1 - 0.83(7.2 - pH)]$$

$$(8.2)$$

where θ is the sludge age to achieve required nitrification (d), K_N the Half-saturation constant for ammonium—at 15 °C this has the value of 0.4 mg/l, $\mu_{N,max}$ the maximum specific growth rate of nitrifiers, typically $0.4 d^{-1}$ at 15 °C, DO the prevalent dissolved oxygen (mg/l) and K_{DO} the half-saturation constant for oxygen (mg/l). The terms K_N and $\mu_{N, max}$ are both temperature dependent and are calculated from

Coefficient	Value (15 °C)
$\mu_m(d^{-1})$	0.4
$K_N(mg/l)$	0.4
$K_{DO}(mg/l)$	1.0
Y (g nitrifiers/g NH_4^+ removed)	0.2

Table 8.2 Kinetic coefficients for the growth rate of nitrifying bacteria in activated sludge.

the relationships:

$$K_{N(T)} = 10^{0.051T - 1.158} \text{ (mg/l)} \qquad (8.3)$$

and

$$\mu_{N,\max(T)} = \mu_{N,\max,(15)} \exp\left[0.098(T - 15)\right](d^{-1}) \quad (8.4)$$

Typical kinetic coefficients for equation (8.2) are given in Table 8.2. This equation is used in a similar way to equation (6.34) and the required sludge age calculated for a given effluent ammonia concentration. The sludge ages required for a range of effluent standards are illustrated in Figure 8.3 and this shows clearly the dramatic effects of temperature on the nitrification process. Thus at temperatures in excess of 20 °C nitrification occurs at sludge ages as low as 3 days, whereas sludge ages in excess of 8 days are required when the temperature falls to 10 °C. In temperate climates with large seasonal temperature differences it is common to specify a summer and winter ammonia effluent standard, and a 2 mg ammonia/l in summer and 5 mg ammonia/l in winter is typical. If nitrification is practised in order to avoid a nitrogenous oxygen demand in the receiving water, then this approach is justified, since the nitrifying bacteria in the receiving watercourse will also be inhibited by the low temperatures.

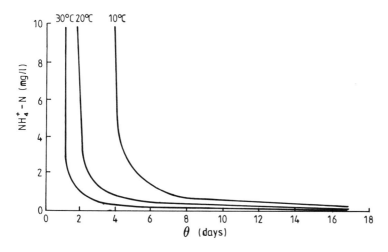

Figure 8.3 The effects of temperature on the sludge age required to produce a given effluent standard for a nitrifying activated sludge.

(b) Application of safety factors

The removal of ammonia from a wastewater has been shown by many workers to follow a zero-order reaction, in other words the rate at which it is removed is independent of the ammonia concentration; it is, however, dependent upon the concentration of nitrifiers in the mixed liquor. Thus if a wastewater treatment plant receives an increased concentration of nitrogen in its influent, the nitrifying bacteria will not respond to this and the excess ammonia will pass out untreated in the effluent stream. In order to compensate for this, the design of nitrifying activated sludge plants incorporates safety factors which permit the build-up of a nitrifying population which can handle changes in influent ammonia concentrations. These safety factors are generally calculated as the ratio of the peak to average TKN loading of the influent to the treatment works, and are applied to the sludge age calculated from equation (8.2). However, it is not usual to apply safety factors in excess of 2.5 as above this value it is unlikely that the required effluent standard will be achieved. In addition, in view of the increased aeration tank volume which results from the application of safety factors, it is likely to be more economic to install flow-balancing tanks to smooth out the wide fluctuations in TKN loadings. The

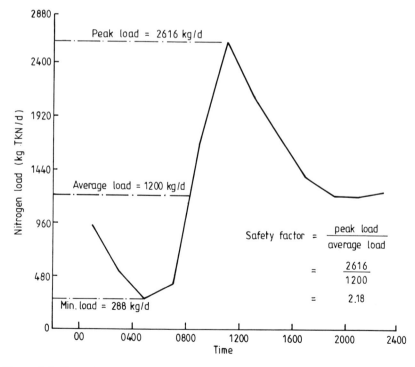

Figure 8.4 Diurnal variations in ammonia loading on a wastewater treatment plant, illustrating the calculation of a safety factor to apply to the sludge age required for nitrification.

calculation of a safety factor for the influent at a typical industrial treatment plant is illustrated in Figure 8.4.

With a value for sludge age corrected using the appropriate safety factor, the required reactor is designed following the procedure outlined in Chapter 7.

(c) Calculation of oxygen requirements

Due to increased oxygen requirements for the nitrogenous oxygen demand, equation (7.40) requires a simple modification. It was demonstrated from equation (8.1) that there is an oxygen demand of 4.2 mg O_2 for each milligram of nitrogen oxidised, therefore expressing the nitrogen as TKN:

$$\text{Nitrogenous oxygen demand (kg } O_2/\text{d)} = 4.2Q(N_0 - N_e) \quad (8.5)$$

where Q is the influent flow rate (m^3/d), N_e the average effluent TKN (kg/m^3) and N_0 the net influent TKN $-$ TKN used in cell synthesis (kg/m^3).

The amount of cells synthesised from 1 g of TKN is calculated from equation (8.1) and was shown to be 0.2 g cells/g TKN. The term N_0 can therefore be written as

$$N_0 = \text{net influent TKN} - 0.2\Delta X \quad (8.6)$$

The total oxygen requirements are calculated from the sum of equations (7.40) and (8.5).

Nitrification using a two-stage activated sludge process

The nitrifying bacteria are slow-growing with a requirement for long sludge ages and high dissolved oxygen concentrations. In addition they are susceptible to inhibition by a wide range of compounds at concentrations so low as not to affect the heterotrophic bacteria. For these reasons it would seem sensible to separate the processes of carbonaceous removal and nitrogen removal into separate reactors, such that different operating conditions can prevail in each, with increased process efficiency and an overall land saving. In addition, it is likely that inhibitory compounds in the influent would be rendered harmless in the first stage by a process of metabolism, dilution and binding to the flocs, consequently nitrification in the second stage would not be inhibited. A schematic diagram of a typical two-stage process is illustrated in Figure 8.5. A wide range of options are available with a two-stage process and a mechanically aerated conventional process has been followed by both a diffused aerator second stage and a nitrifying trickling filter. In addition the first stage can also vary between pure oxygen systems such as the VITOX process, mechanical aerators and diffused air. In view of the wide

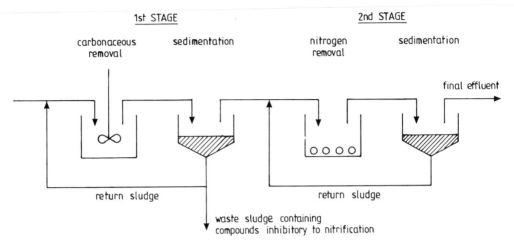

Figure 8.5 Schematic diagram of a two-stage nitrifying activated sludge plant. Stage 1 is a mechanically aerated carbonaceous removal process, whereas stage 2 is a diffused air nitrification process.

range of combinations and the fact that very few two-stage plants are in operation, very little design information is available. A common problem of two-stage systems is that due to the low growth yield of the nitrifiers the solids concentration in the second-stage reactor is very low. This frequently leads to poor sludge settleability with attendant loss of solids. Frequently the only way to alleviate this is to recirculate solids from the first-stage sedimentation tank. In addition, diffused air systems operating at a low solids concentration are associated with foaming problems, and antifoam addition is frequently required. Finally, in view of the reduction of wastewater alkalinity which occurs in the first stage together with the enhanced nitrification in the second stage, there is a large drop in the pH of the second-stage reactor which is often large enough to need pH control.

Nitrification in trickling filters

In a single-stage trickling filter, the nitrifying bacteria will be competing with the heterotrophic bacteria for their oxygen supply. The availability of oxygen within a filter is a function of the BOD concentration and the heterotrophic bacteria will outgrow the nitrifiers when BOD is readily available. It appears that a soluble BOD of as low as 20 mg/l is required before sufficient oxygen is available to permit nitrification. As very few filters are capable of producing this quality effluent, nitrification is frequently absent or confined to the lower reaches of the filter. In order to achieve consistent nitrification by trickling filters it is necessary to limit the organic loading rate; for mineral media a figure of 0.16–0.19 kg/m³ d

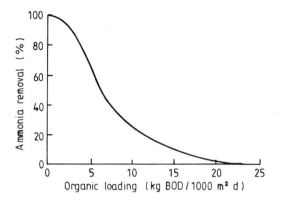

Figure 8.6 The effect of organic loading rate on ammonia removal in trickling filters.

is often quoted to achieve an ammonia removal of 75%. At loading rates above $0.4\,kg/m^3\,d$ very little nitrification is obtained. A typical relationship between loading rates and ammonia removal rate is shown in Figure 8.6, and this can be used to select the filter area required in order to achieve a given effluent ammonia concentration.

Because of the large reserve of biomass associated with a trickling filter, they are less susceptible to load variations than suspended growth processes, and twofold variations in ammonia loadings do not affect the effluent quality. They are, however, very susceptible to flow variations, and a flow sufficient to keep the media continuously wet must be provided.

8.3 DENITRIFICATION

Introduction

In the absence of a supply of dissolved oxygen, the utilisation of oxygen as the terminal electron acceptor for respiration is inhibited. Under these conditions the majority of facultative microorganisms have to rely on fermentation in order to regenerate NAD^+ (Chapter 5). However, certain chemoorganotrophs are capable of replacing O_2 with NO_3^- as the terminal electron acceptor and respiration can proceed with the reduction of nitrate to nitrite, nitric oxide, nitrous oxide or nitrogen as shown in equation (8.7).

$$
\begin{array}{cccccc}
\text{Redox} & +5 & +3 & +2 & +1 & 0 \\
\text{state of} & NO_3^- & \rightarrow NO_2^- & \rightarrow NO & \rightarrow N_2O & \rightarrow N_2 \\
\text{nitrogen} & \text{Nitrate} & \text{Nitrite} & \text{Nitric*} & \text{Nitrous*} & \text{Nitrogen*} \\
& & & \text{oxide} & \text{oxide} &
\end{array}
$$

$$(8.7)$$

where the asterisk denotes gaseous end-products. This process is known as anaerobic or nitrate respiration and is carried out by a variety of bacteria such as *Alcaligenes, Achromobacter, Micrococcus* and *Pseudomonas*. Not all these genera are capable of complete oxidation to nitrogen and a variety of gaseous products can be produced.

The redox state of the intermediates in denitrification (reaction 8.7), show that the reaction proceeds in a series of steps, each one associated with the gain of one electron. An electron donor is therefore required as a source of these electrons. In sewage treatment, this reaction is carried out primarily by heterotrophic bacteria and only organic carbon sources can be used. Although the wastewater itself contains a suitable source of organic carbon this is inadequate in treated effluents, thus in two-stage systems a supplemental source of carbon must be provided. This is frequently achieved by using industrial and agricultural wastes such as brewery waste, molasses or corn-silage. In the absence of such an alternative, methanol is generally accepted as the most appropriate commercially available carbon source. The stoichiometry of growth on methanol as both a carbon and energy source is given by

$$NO_3^- + 1.08CH_3OH + H^+ \rightarrow 0.065C_5H_7O_2N + 0.47N_2 + 0.76CO_2 \\ + 2.44H_2O \qquad (8.8)$$

This reaction highlights the differences between the growth of the nitrifying bacteria (equation 8.1) and the denitrifiers. Oxygen is not required for denitrification; indeed, when it is present, it is preferentially exploited as a terminal electron acceptor. In addition, as these organisms utilise protons in the reduction of nitrate, then the wastewater will tend to become alkaline compared with the acidity produced during nitrification. Finally, as the denitrifiers are heterotrophic bacteria, they are energetically much more efficient than the nitrifiers and thus their yield and growth rate will be much higher.

Kinetics of denitrification

The two important factors which affect the rate of denitrification, are the substrate (electron donor) concentration and the nitrate concentration. Both these effects can be modelled using standard Monod kinetics, and the growth rate is described by a double Monod equation of the form:

$$\frac{d(NO_3)}{dt} = \mu_m \frac{N}{K_N + N} X \qquad (8.9)$$

where μ_m is the maximum specific growth rate of the denitrifying bacteria and N the nitrate concentration.

Values for the nitrate saturation coefficient (K_N) are generally very low, in the range 0.08–0.1 g/l and thus $N \gg K_N$ and the Monod term for nitrate concentration in equation (8.9) approximates to 1. Therefore this equation can be written as

$$\frac{d(NO_3)}{dt} = \mu_m X \qquad (8.10)$$

This implies that denitrification is a first-order reaction with respect to biomass concentration and zero order with respect to nitrate concentration. For a complete mix reactor of volume V, a mass-balance equation of the type described by equation (7.9) can now be constructed:

$$\left(\frac{d(NO_3)}{dt}\right)V = QN_0 + Q_rN_e - (\mu_m X)V - (Q + Q_r)N_e \qquad (8.11)$$

At steady state when $d(NO_3)/dt = 0$, then:

$$\frac{Q(N_0 - N_e)}{VX} = \mu_m \qquad (8.12)$$

In this case the term μ_m is often referred to as the specific denitrification rate, $(q)_{DN}$. It is related to temperature by the empirical equation:

$$(q)_{DN} = 0.07(1.06)^{T-20} \qquad (8.13)$$

Thus assuming that the temperature of the coldest month is known, a reactor volume is calculated from equation (8.12) once an operating mixed liquor solids concentration has been selected.

Process configurations for nitrogen removal

The complete removal of nitrogen from a wastewater requires that both nitrification and denitrification occur, as denitrification cannot proceed without the presence of nitrate. As the two reactions appear to have fundamentally opposite environmental requirements, particularly with respect to oxygen, then it is difficult to see how they can both occur within a single reactor. However, by providing zones within the reactor where aerators are switched off, and only mixing occurs, anoxic conditions are quickly established and denitrification will occur. This can be achieved most easily in plug-flow reactors, baffled tanks or endless channel oxidation ditches. The anoxic zone is generally chosen close to the point where the settled sewage and return sludge are fed to the reactor to ensure that there is an adequate supply of electron donors in the settled sewage, and nitrate via the recycled sludge. In order to ensure that the effluent nitrate concentrations are met, it is necessary to recycle a large fraction of the sludge and a recycle ratio of 1.5–2 is often required. After the anoxic period, aeration commences and nitrification is quickly resumed. A flow diagram for a typical single-sludge nitrogen removal reactor is shown in Figure 8.7.

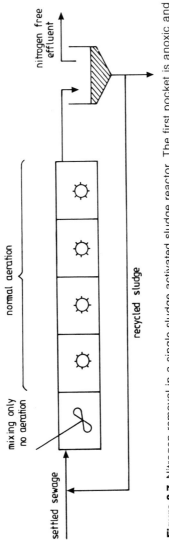

Figure 8.7 Nitrogen removal in a single-sludge activated sludge reactor. The first pocket is anoxic and receives the return sludge and settled sewage. The remaining pockets are aerobic and operated at a long sludge age in order to ensure full nitrification.

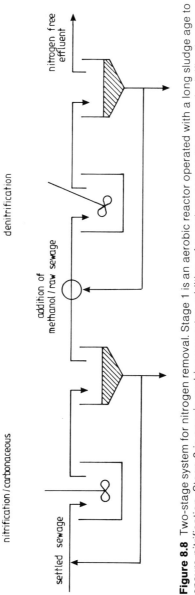

Figure 8.8 Two-stage system for nitrogen removal. Stage 1 is an aerobic reactor operated with a long sludge age to ensure nitrification. Stage 2 is anoxic and incorporates addition of a carbon source to ensure denitrification.

A higher nitrogen removal rate can generally be achieved in a separate sludge system in which the effluent from the nitrification stage, which is high in nitrate, is fed to a separate anoxic reactor for denitrification (Figure 8.8). The higher removal rate means that a lower overall reactor volume is required, but provision of two clarifiers means that the clarification requirements are increased. In addition, as the effluent from the first stage is fully nitrified, it will also have a low BOD and consequently be deficient in electron donors, an additional carbon source is therefore required. Finally the high rates of denitrification often result in an increased pH and thus pH control must be provided. Single-sludge systems are therefore generally more cost effective and require less process control.

8.4 PHOSPHORUS REMOVAL

Introduction

Of the nutrients which are capable of supporting luxuriant growths of algae in a receiving water, phosphorus is rate limiting, and a concentration of $10 \mu g/l$ is required before algal growth will occur. It has been argued, therefore, that control over phosphorus-containing compounds in aquatic ecosystems presents a means of controlling the deleterious effects of eutrophication. Consequently, if the small concentration of phosphorus which is present in sewage effluents can be removed, then algae will not be able to flourish, regardless of the nitrogen concentration. Phosphorus load control has been demonstrated as one of the most effective ways of dealing with man-made eutrophication; for this reason several countries apply a phosphorus standard for sewage effluent discharges, as well as one for nitrogen. A typical phosphorus standard is 1 mg/l dissolved ortho-phosphate (as phosphorus). The major sources of phosphorus in domestic wastewaters are from human excreta (50–65%) and synthetic detergents (30–50%) and typical concentrations are in the range 10–30 mg/l as phosphorus.

Enhanced biological phosphorus removed in activated sludge systems

The uptake and removal of phosphorus from a wastewater by activated sludge is a relatively new development which followed from the observation that if an activated sludge is allowed to become anaerobic, then the amount of phosphorus (as phosphate, PO_4^{3-}), in the supernatant increases. Upon resumption of aeration, however, there is a rapid uptake of phosphate by the sludge which is in excess of that released during anaerobiosis (Figure 8.9). This 'luxury uptake' of

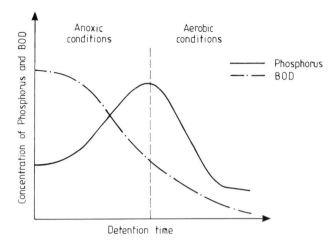

Figure 8.9 Enhanced phosphorus uptake observed in activated sludges when transferred from anoxic to aerobic conditions.

phosphate results in a phosphate-depleted mixed liquor and a phosphate-rich sludge. The ability to release or store phosphorus in a sludge by manipulating the prevailing oxygen concentration, has been exploited in several process, as will be seen later in this chapter.

The mechanism by which phosphorus is stored and released is extremely interesting and helps in understanding the operation of phosphorus-removal activated sludge plants. Several studies have shown that the removal and release of phosphorus within a sludge is the result of a single genus of bacteria known as *Acinetobacter* spp. and more specifically a single species, *Acinetobacter calcoaceticus*, is implicated. The *Acinetobacter* spp. are relatively easily isolated from a range of environments including soils and sewage. They can utilise sugars as a source of carbon and energy and these are degraded by the Entner–Doudoroff pathway (p. 153). As this is operative under aerobic conditions, they are only able to utilise sugars under aerobic conditions. Under anaerobic conditions, they are capable of degrading volatile fatty acids, in particular acetate. *Acinetobacter* bacteria also possess two storage polymers: poly 2-hydroxy butyrate (phb) is an electron sink for the storage of excess organic carbon when other nutrients are limiting; volutin or metachromate are granules of polyphosphate formed from phosphate according to the reaction:

$$ATP + (PO_4)_n \xrightleftharpoons[\text{Degradation}]{\text{Synthesis}} ADP + (PO_4)_{n+1} \qquad (8.14)$$

An alternative degradative route is by hydrolysis:

$$H_2O + (PO_4)_n \longrightarrow (PO_4)_{n-1} + HPO_4^{2-} + H^+ \qquad (8.15)$$

Thus degradation of polyphosphate is associated with energy release, and *Acinetobacter* spp. subjected to anaerobiosis, are capable of taking up acetate and using this to synthesise phb (Figure 8.10). The energy for phb synthesis comes from the degradation of

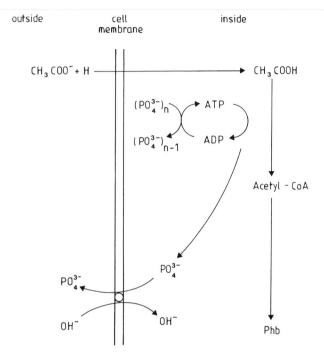

Figure 8.10 The uptake of acetate by *Acinetobacter* under anaerobic conditions and the synthesis of phb, concomitant with polyphosphate degradation and the release of phosphate across the cell wall.

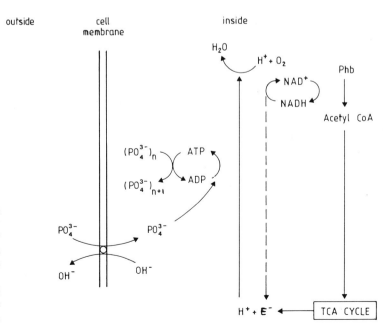

Figure 8.11 Phosphate uptake by *Acinetobacter* under aerobic conditions which is used in the synthesis of ATP and polyphosphate, using the energy from phb degradation.

polyphosphate, and there is a consequent increase in the mixed liquor phosphate concentration. Under anaerobic conditions, acetate is generally present in wastewaters as it is an end-product of metabolism for many facultatives anaerobic bacteria (volatile fatty acid formers, see p. 156).

When the *Acinetobacter* pass to an aerobic environment they are surrounded by a mixed liquor which is low in organic substrate, as the readily available organics have been removed in the anaerobic phase. However, the degradation of stored phb to acetate will provide both a carbon and energy source, with the generation of ATP. As a result of the availability of ATP, polyphosphate is removed from the bulk solution into the *Acinetobacter* granules (Figure 8.11). Such a mode of life makes the *Acinetobacter* spp. a very effective competitor for limited food supplies. Up to 40% of the viable fraction of bacteria in a conventionally activated sludge plant have been identified as *Acinetobacter* spp. This rises to 80% for enhanced biological phosphorus removal plants.

8.5 SPECIALISED TREATMENT SYSTEMS FOR ENHANCED PHOSPHORUS REMOVAL

The Phostrip process

One of the earliest application of 'luxury uptake' was the Phostrip process. This maintains an aerator and clarifier under typical activated sludge operating conditions, but instead of sludge being wasted from the clarifier it is fed to a separate unstirred tank. This tank is anaerobic and the sludge is retained for several hours, during

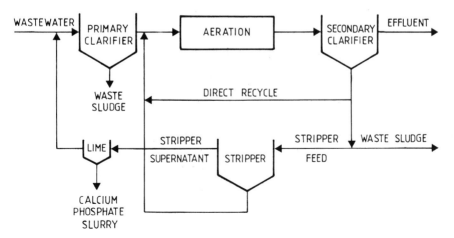

Figure 8.12 Schematic diagram of the Phostrip process for the removal of phosphorus in the form of a calcium phosphate slurry.

which phosphate is released. The phosphate-rich supernatant is then fed to a tank which is dosed with lime so that phosphate is precipitated according to the reaction:

$$3Ca(OH)_2 + 2PO_4^{3-} + 6H^+ \underset{<9.0}{\overset{>9.0}{\rightleftharpoons}} Ca_3(PO_4)_2 + 6H_2O \quad (8.16)$$

The phosphate-depleted sludge is recycled back to the aeration basin where luxury uptake of phosphorus occurs (Figure 8.12). The excess phosphorus is thus removed as a calcium phosphate slurry.

The Bardenpho process

This nutrient-removal activated sludge process removes both nitrogen and phosphorus from a wastewater as well as achieving an extremely high carbonaceous BOD removal. It is based around a modified activated sludge process, with a number of unit operations appended each with a separate function. With reference to Figure 8.13, the process is based around an aeration basin (1) which is designed and operated to give a fully nitrified effluent. This is recycled to a basin ahead of the influent aerator (2) at a recycle rate of five times the influent flow. Here denitrification takes place and 80% of the nitrates are removed. That flow which is not recycled is passed to a second anoxic basin (3) where the remaining nitrates are removed. This nitrogen-free effluent is now aerated (4) before passing to the clarifiers. The sludge from the clarifier is returned to an anaerobic reactor where phosphate release occurs (5), thus on passage to the aerobic zone, luxury uptake of phosphorus occurs. The phosphorus is thus removed from the process as a phosphate-rich sludge and the nitrogen is removed as nitrogen gas. A comparison of the performance of a typical Bardenpho plant with the conventional activated sludge process is shown in Table 8.3. Obviously this process is a

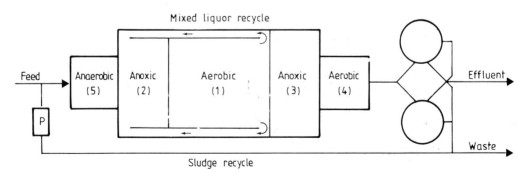

Figure 8.13 The Bardenpho process for the removal of phosphorus in the form of a phosphate-rich sludge.

Parameter (mg/l)	Conventional	Bardenpho
BOD$_5$	20	< 5
COD	80–120	15–40
Total nitrogen	50–70	1–2
Nitrate	< 1	2–3
Ammonia	15–30	< 0.5
Suspended solids	30	< 10
Phosphorus	15–25	0.5–1

Table 8.3 Comparison of final effluents from a Bardenpho and a conventional, non-nitrifying activated sludge plant.

highly sophisticated modification of the activated sludge process which produces a high-quality effluent. There are only a limited number of circumstances and environments for which it would be necessary, and in view of the sophisticated control required it is quite unsuitable for use in developing countries.

9 Operation and Control of Conventional Systems

9.1 TRICKLING FILTERS

Introduction

Trickling filter systems are not complicated and when problems are encountered the number of options available for dealing with them are limited. The best way to alleviate problems is to ensure regular and planned inspection and maintenance. The majority of problems with trickling filters arise from neglect and the remainder are due to overloading. If overloading does occur it is difficult to alleviate without incurring additional operating costs.

Filter overloading

When the design hydraulic and organic loading to a filter is exceeded and this results in a deterioration of effluent quality, a number of strategies are available, based on altering the flow regime to the filters. A single-stage filter generally has a loading of $0.12\,kg\ BOD/m^3\,d$, which for most domestic sewages gives a hydraulic loading of $0.5\,m^3/m^3$. If this load is exceeded then effluent quality deteriorates and the filter will pond. If the overloading is severe then ponding can cause a complete breakdown in treatment. This situation can in many cases be alleviated by recirculation of final effluent; the mode of recirculation employed will depend upon the size and flexibility of the works. There are two major types of recirculation given under the heading below.

1. Single-stage filtration with recirculation

The most common type of recirculation involves recycling a fraction of the treated effluent from the clarifier and mixing this with the feed to the filter, generally after the primary sedimentation stage (Figure 9.1). The effect of this is to dilute the concentration of BOD of the filter influent. In addition it provides a more uniform hydraulic load by evening out the large diurnal fluctuations in flow and also provides a more effective wetting of the media. The effect of diluting the influent feed strength is to reduce the amount of ponding by causing a reduction in microbial film thickness. The increased hydraulic loading also has the effect of pushing the biofilm further down the filter bed and therefore reducing the film thickness at the surface.

The amount of recirculation required will depend on the influent and effluent organic strength and recirculation rates from one to five are common. Design charts such as Figure 7.8 can be used to calculate the necessary recirculation, but the results should be viewed with caution because of the other variables which are changed by recirculation. The most important of these is the specific surface area of the media, since the increased recirculation causes an increased wetting of the media, then the effective surface area is increased. Where the filter overloading is very high it is more beneficial for the receiving water if the whole of the flow is passed through the filters, even if BOD removal is poor. The discharge of a portion of the flow as untreated sewage should be avoided. The logic for this is based on the nature of the organic material in wastewater. A portion of the BOD is readily biodegradable and is removed rapidly by the filters, the remainder is less readily degradable, and in overloaded filters this fraction is discharged to the river. Because it is biodegraded more slowly, the oxygen demand it exerts will be dispersed over a much greater area of the receiving water and thus its detrimental effect reduced. Recirculation permits increases in organic loading up to $0.15 \, \text{kg BOD/m}^3 \, \text{d}$ and hydraulic loadings up to $0.9 \, \text{m}^3/\text{m}^3 \, \text{d}$.

2. Two-stage alternating double filtration

This technique was suggested in 1922 following the observation that ponding on filters could be alleviated by application of final effluents from activated sludge plants. This concept was then advanced by using treated effluents from trickling filters and similar results were obtained. It is now exploited using a pair of filters which act alternately as the primary and secondary filters. The primary filter receives settled sewage at a high loading rate which will ultimately lead to ponding, the secondary filter receives the effluent from this filter and it receives a low loading because of the BOD removed by the first filter. The loading to the filters is reversed periodically and the secondary filter receives the settled sewage (Figure 9.1). Because the primary filter is now receiving a much-reduced organic loading there is insufficient nutrient to support the heavy growth of biofilm, and this dies and sloughs off. The key to successful alternate double filtration is the length between change-over periods. This must be carried out before ponding occurs on the primary filter and it is highly dependent upon the season. A change-over period from 6 days up to 2 weeks is the most common. A well-operated alternate double-filtration filter

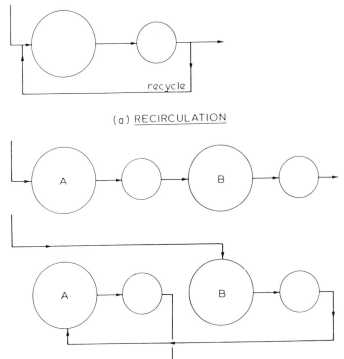

(a) RECIRCULATION

(b) ALTERNATING DOUBLE FILTRATION

Figure 9.1 Flow diagram for trickling filters incorporating: (a) recirculation; (b) alternate double filtration.

can handle an organic loading of up to $0.25\,kg\,BOD/m^3\,d$ and a hydraulic load of $1\,m^3/m^3\,d$. As well as the additional pumping costs, alternate double filtration requires the provision of additional humus tanks and at many works it may not be possible to provide the required arrangement of piping.

Fly problems

All filters are extensively colonised by fly larvae, and although more than 30 fly species are routinely encountered in filter media the two species *Psychoda* and *Sylvicola* appear to cause the most problems. Chironomid midges are also abundant in underloaded filters. The problem of emergent flies is due mainly to their emergence from filters being synchronised, which can result in heavy swarms. Their frequency of emergence increases as the temperature increases and so they pose more of a problem in hot countries than temperate ones. Although fly problems can be alleviated by dosing the filters with pesticides, this is unwise due to potential problems of their persistence in aquatic food chains. The problem is particularly actute in developing countries where pesticides such as DDT are still available, despite their use being prohibited in the UK and USA. Fly problems can be reduced by correct operating procedures, and since fly larvae feed by grazing on the biofilm, heavy film growth will encourage heavy larval growth. Heavily loaded single-pass filters are most often responsible for fly problems and these are reduced by introduction of recirculation. The method of sewage application can also influence the filter fauna. Distributor arms fitted with splash plates give a more even distribution of sewage and this has been found to reduce the numbers of *Psychoda* and *Sylvicola*. Conversely, where the distribution system produces uneven wetting, the drier areas of the filter medium provide greater densities of chironomid larvae.

9.2 ACTIVATED SLUDGE PLANTS

Introduction

Activated sludge plants are designed to produce a specified effluent quality and the aim of plant operation and control is to ensure that the effluent quality is maintained at the lowest operating cost. Wide variations in effluent quality are a feature of most plants and the nature and extent of this variability is determined by a number of factors. The most important of these factors are:

1. Variations in the flow rate, BOD and suspended solids concentration in the influent to the treatment plant;

2. Fluctuations in plant operational parameters, in particular sludge age, F/M ratio, mixed liquor solids concentration, recycle ratio and solids concentration, and the solids wastage rate;

3. Fluctuations in the oxygen demand of the waste which are not matched by the aeration regime;

4. Settling characteristics of the sludge in terms of the stirred sludge volume index;

5. Environmental factors such as temperature and wind.

The activated sludge process performs more efficiently if it is operated at a steady state, and an understanding of how process variability affects system performance helps in plant operation, leading to decreased effluent variability.

Influent variables

The nature of the variations in influent loading are a function of the sewerage system which serves the treatment plant and the dischargers who use the sewerage system. Many industries operate batch processes while others have seasonal flows, and these will cause a step change in input flow and concentration. Heavy rainfall has a similar effect in that it causes a sudden change in influent flow. Diurnal, weekly and seasonal changes in flow and load are examples of periodic influent variations. The magnitude of the range of these flow and load variations are inversely proportional to the population served by the wastewater collection system. Large systems with long retention times provide a considerable dispersion of peak flows and loadings, whereas in small systems the retention time is too short for any dispersion.

Wide variations in influent flow generally prove more detrimental to effluent quality than variations in influent strength. Increased influent flows cause decreased hydraulic retention times and a hydraulic surge through the plant. This produces a higher overflow rate in the final clarifier which can often lead to the wash-out of solids over the effluent weir. Plants which are subjected to step changes in influent flow associated with detrimental effluent quality can frequently reduce these through trade effluent control, or alternatively by the installation of flow-balancing tanks.

The equations derived in Chapter 7 from steady-state mass-balance models show that effluent organic concentration is independent of the influent organic concentration. This has been confirmed by a number of investigations for a range of loadings, but it is not valid for high- and variable-strength influents (usually industrial wastes). For such wastes an equation has been presented which allows the calculation of the required organic loading to maintain a consistent effluent:

$$S_e = \frac{S_0}{(KF^{-1}) + 1} \qquad (9.1)$$

where S_e is the required effluent quality (mg COD/l), S_0 the peak influent waste strength (mg COD/l), K the specific substrate removal rate coefficient (d^{-1}) and F the organic loading (kg COD applied/kg MLSS d).

The implications of this equation are that, for a constant organic loading rate, variations in influent organic strength will produce variations in effluent quality. Assuming that time-based variations in loading were considered at the design stage, and adequate flexibility exists in the treatment system, these variations can be reduced by appropriate reductions to the organic loading rate.

Plant operational parameters

The two most important parameters which a plant operator can vary in order to minimise effluent variability are the rate of return of

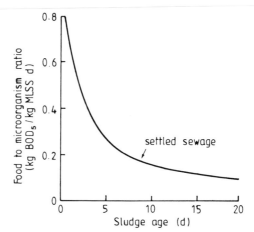

Figure 9.2 The effects on the F/M ratio of varying the sludge age for a plant treating settled domestic sewage.

activated sludge from the clarifier to the aeration tank and the rate of wasting of excess sludge. Return of activated sludge provides sufficient concentration of biomass in the aeration tank, and wasting excess sludge is required to maintain either a given F/M ratio or sludge age. It is not possible to maintain both a constant F/M and a constant sludge age (Figure 9.2) and operators must select one or the other as a control parameter.

1. Sludge age

Selection of the appropriate sludge age for a particular plant depends upon the required effluent quality and is calculated from equation (6.17). There is a wide operating region of sludge ages which produce only small changes in effluent quality. A sludge age of at least 3 days is required, and below this, sludge settleability is poor (Figure 9.3), although the exact form of this relationship is plant and wastewater specific. At sludge ages greater than 10 days, defloccul-

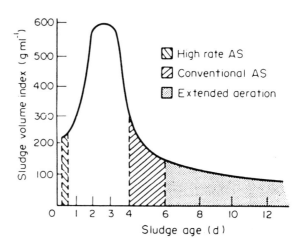

Figure 9.3 The influence of sludge age on the settling properties of the sludge.

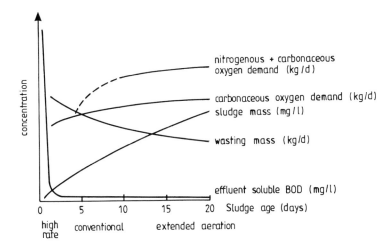

Figure 9.4 The effects of increasing sludge age on plant performance and oxygen demand.

ation of the sludge flocs occurs and there is an increased effluent turbidity. A number of parameters vary with increasing sludge age of which the most significant is the onset of nitrification. This occurs at sludge ages in excess of 6 days in temperate climates, and at much lower sludge ages in hot climates. Other factors affected by changes in sludge age are illustrated in Figure 9.4.

Sludge age is generally controlled by varying the wastage rate of excess solids from the process. Sludge can either be wasted as thickened sludge from the recycle line, or as MLSS from the aeration basin. The former is the most common as it means that a much smaller volume of sludge has to be wasted and thus pumping and disposal charges are reduced. Wasting is generally performed on a batch basis and the amount of sludge to be wasted is calculated from equation (7.12). A major criticism of this method is the delay which occurs between samples of mixed liquor and recycle sludge being taken, analysed and the results conveyed to the plant operator. As a result of diurnal variations in recycle sludge concentration, this time delay may mean that the amount of sludge actually wasted may be very different to that intended. This is circumvented by taking composited samples of the waste sludge during the waste period, to confirm that the required amount of sludge has been wasted. As this rarely happens, the actual operating sludge age of a plant can be very different from that which the operator believes it to be.

An alternative method of controlling sludge age is to waste excess solids directly from the aeration basin. This is known as hydraulic control and determination of the amount of MLSS to waste is very simple. Assuming that the volume of mixed liquor abstracted from the reactor each day is v, and the required sludge age is calculated from equation (7.41), then:

$$\theta = \frac{VX}{\Delta X} = \frac{VX}{vX} = \frac{V}{v} \qquad (9.2)$$

Thus the sludge age is fixed by the fraction of mixed liquor wasted from the aeration basin. The above equation is not corrected for solids lost in the final effluent and these must be taken into account by reducing the wastage rate accordingly. Hydraulic control of sludge age incurs additional construction costs as it requires a separate sedimentation tank to settle out and thicken the solids which are to be wasted. It does, however, guarantee a constant sludge age, and takes the onus for this out of the plant operator's hands; however, regular calibration of associated pumps is essential. The conventional secondary sedimentation tank is used for clarification of the final effluent before discharge, and all the thickened solids from this tank are returned to the aeration basin.

2. Food:microorganism ratio

The F/M ratio is an alternative control parameter to sludge age and one which is widely used in practice. Because it relies on BOD_5 values to calculate the influent organic loading it is largely a historical control parameter. Knowing the average weekly flow and load to a plant the loading rate is fixed by controlling the mixed liquor solids concentration. The mixed liquor solids concentrations required to give a range of loading rates are illustrated in Figure 9.5, which also shows that for a fixed F/M ratio the sludge age is dependent upon the mixed liquor solids concentration. The popularity of the F/M ratio as a control parameter is due largely to the apparent ease with which it is controlled, that is, by maintaining a fixed concentration of mixed liquor solids in the aeration basin. In a similar way to sludge age, the settleability of the sludge appears to be very dependent upon the F/M

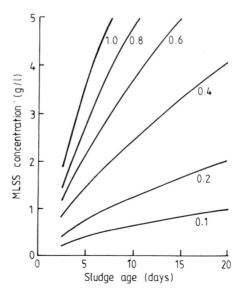

Figure 9.5 The influence of the mixed liquor solids concentration on the sludge age at a fixed F/M ratio over the range 0.1–1.0 kg BOD/kg MLSS d.

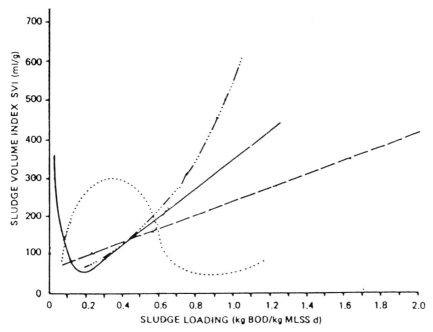

Figure 9.6 Typical relationships which have been established at different plants for the effects of variations in the F/M ratio on the sludge-settling properties.

ratio (Figure 9.6) although this relationship appears to be highly plant specific. The major criticisms of the F/M ratio are that it produces neither a constant loading rate nor sludge age, since if the MLSS are constant the actual loading will vary diurnally depending upon the variations in flow and load to the plant. In addition, although the F/M sets the MLSS value which must be attained, it gives no indication of the sludge mass which must be wasted in order to achieve this value.

Operation and control of aerators

The process of aeration utilises significant amounts of energy and for some plants this may be up to 80% of the total energy budget. In terms of overall annual expenditure, however, energy costs can be as low as 8% of the total annual costs and rarely greater than 25% (Figure 7.4). Thus although potential monetary savings are large at larger works, these will be small when compared to total annual costs. Although it is essential that the process of aeration is optimised, capital expenditure to improve operational flexibility is rarely justified. This means that potential energy savings must be realised by operational means. Before embarking on energy-saving exercises it is important to estimate the amount of energy which it is possible to save, assuming that the plant is operating under optimum conditions. This is achieved by comparing the actual energy used for aeration with the theoretical energy requirements of the process. Theoretical energy requirements are found from the aerator efficiency and the oxygen demand under various sludge loadings. It is reasonable to expect an answer within 10% of the theoretical daily oxygen requirement. The minimum energy requirement will be fixed by the sludge loading to the plant and the required effluent quality. The quality of final effluent should never be compromised in order to make energy savings.

If a desk-top study indicates that energy savings are possible then the two most important cost-saving exercises are as follows:

1. To ensure that the oxygen supply matches the oxygen demand at all times and at all locations in the aeration basin;

2. Where nitrification is practised, the oxygen held in the nitrate ion is released by denitrification, by incorporation of an anoxic zone.

In order to match the oxygen supply to oxygen demand, it is necessary to know the oxygen profile across the aeration tanks. Although this can be measured relatively easily for completely mixed plants, it proves more difficult in plug-flow reactors. These often experience oxygen deficiency at the head of the reactor, but experience oxygen excess towards the outflow. A number of methods are available for matching energy input to the oxygen demand profile and these will depend upon the aerator type. Diffused aerators frequently have a number of blowers which cut in or in out depending upon the oxygen demand; alternatively the blowers have vanes of variable speed. The power output of fixed-speed surface aerators is dependent upon the depth of immersion of the aerators. This can be varied either by variable-height weirs which provide a variable depth of mixed liquor in the aeration basin, or a variable aerator which has the immersion depth varied to match the oxygen demand. A third alternative is to use variable-speed aerators which have two operating speeds, the speed is then selected to meet the oxygen needs of the influent. The simplest technique of all, however, is to switch off one or more aerators in the aeration tank as the load to the works declines. It has been shown that this approach rarely causes the solids to flocculate and settle in the reactor pocket without aeration. If the flexibility is not available to vary the aeration regime, an alternative approach is to vary the pollution load for treatment by flow balancing, such that peaks and troughs in organic loadings are reduced.

9.3 SLUDGE SETTLEABILITY AND OPERATION OF CLARIFIERS

The major source of variability in effluent quality is due to loss of suspended solids in the final effluent. As these solids often constitute

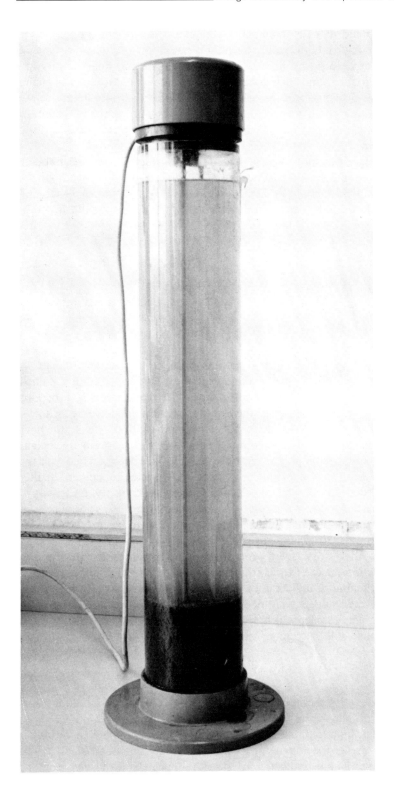

Figure 9.7 Apparatus used in the determination of SSVI.

the majority of organic material in effluents, efficient clarification is essential if a high-quality, stable effluent of low BOD and suspended solids is to be produced. The aeration tank and final clarifier are interrelated and their functions interdependent. Although in many cases efficient operation of secondary clarifiers is the limiting factor in producing a high-quality effluent, the settling properties of the sludge are determined primarily by the conditions prevalent in the aeration basin.

The settleability of an activated sludge is quantified by means of the stirred sludge volume index (SSVI) which is also known as the stirred specific volume (SSV). This is a measure of the volume occupied by 1 g of sludge (ml/g) and it is determined in the apparatus illustrated in Figure 9.7. The sludge is stirred at 1 rpm and settled for 30 min, after which time the height of the interface is noted. The SSV is then calculated as

$$SSV(ml/g) = \frac{\text{Interface height after 30 min}}{\text{Initial height of sludge} \times \text{Suspended solids (mg/l)}} \times 10^6$$

(9.3)

This parameter is used in the design of secondary sedimentation tanks and it determines the maximum solids loading which can be applied to a sedimentation tank. Most plants suffer sporadic incidents of poor sludge settleability when the SSV increases from an average value of < 80 to > 120. The sludge is then said to have 'bulked' and the flocs have a low density and compact poorly. When this occurs the height of the sludge blanket rises, with a risk of solids being carried over the weir with the final effluent. The effect of bulking on effluent quality will depend on available sedimentation tank capacity; if capacity is generous poor settleability rarely causes problems. Similarly, if bulking is not too severe the effects may only be noticed at peak loadings. To maximise use of settlement tanks it is important to know how they work (see Chapter 3). Sedimentation tanks perform two functions, namely the clarification of an effluent for discharge and the thickening of a sludge for recycle and wasting. With concentrated activated sludges the thickening function limits maximum through-put. The rate at which solids are applied to the tank is called the applied solids loading (ASL), given by

$$ASL = \frac{(Q_i + Q_u)MLSS}{A}(kg/m^2\,h)$$ (9.4)

where Q_i is the influent flow rate (m³/d), Q_u the recycle sludge flow (m³/d) and A the settlement tank cross-sectional area (m²).

The limit to the amount of solids which can be applied to the tank is known as the maximum permissible solids loading (MPSL). The faster a sludge is able to settle, the faster it can be withdrawn from the

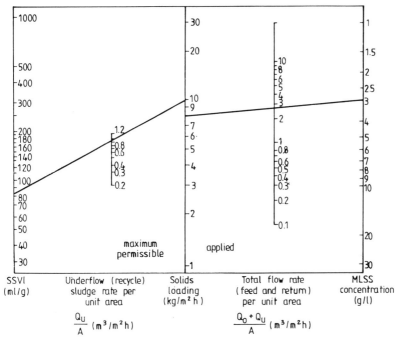

Figure 9.8 Nomograph prepared by the Water Research Centre for efficient operation of secondary sedimentation tanks (From White, 1975. Reproduced by permission of the Water Research Centre.)

tank and the higher will be the MPSL. If the ASL does not exceed the MPSL then the sludge blanket will remain at a constant level, but if it is exceeded then the blanket will rise until it reaches the outlet weir, when a loss of solids will occur. The ASL can be calculated from equation (9.4) and obviously it will vary throughout the day. The MPLS is dependent on the recycle rate and the sludge settleability, and a nomograph is used to simplify its calculation (Figure 9.8). When a severe incidence of bulking occurs, the nomograph can be used to ensure that the sedimentation tanks are operating at their most efficient. It is used as follows:

1. Measure the SSV of the sludge and mark this on the chart.

2. Measure the recycle sludge flow rate and the maximum influent flow rate.

3. Calculate the MPSL from the nomograph.

4. If MPSL > ASL use the chart to calculate a recycle ratio which will give an ASL of MPSL/1.2.

5. If ASL > MPSL use chart to calculate a new increased recycle rate. Note that recycle rate should not exceed 1.2 m³/m² h.

6. If the recycle rate exceeds $1.2 \, m^3/m^2$ h then the wastage rate must be increased to reduce the MLSS. These should not drop below 2000 mg/l.

7. If the above criteria cannot be met, then the secondary sedimentation tank is overloaded and steps must be taken to find the cause of bulking and eliminate it.

Settlement and bulking of activated sludges

Under the quiescent conditions of the secondary sedimentation tank, an activated sludge will settle out and thicken with up to a fourfold increase in solids concentration. This settling ability is a direct result of the ability of the microorganisms and colloidal material which comprises the reactor mixed liquor solids to agglomerate into large flocs with a high zone-settling velocity. This agglomeration process is known as flocculation and it is essential if efficient sewage treatment is to be achieved.

The loss of settleability, which is characteristic of a bulking sludge, means that the sludge-settling rate decreases and thickening is poor. This leads to the production of dilute return and activated sludge streams, hydraulic overloading of solids handling processes, high effluent solid and BOD levels and, in severe cases, wash-out of the activated sludge reactor. At many activated sludge plants, thickened sludge is often dewatered in order to reduce handling costs. Flocculation is very important in determining the dewatering characteristics of a sludge. Sludge dewatering characteristics are influenced by biological-induced flocculation and a good flocculating sludge will dewater easily. Although sludge volume is less than 1% of the total plant influent, sludge handling may account for as much as 60% of the total capital and operating costs.

Two mechanisms have been proposed to explain sludge bulking; one suggests that bulking results from an increased floc surface charge brought about by changes in floc surface chemistry, the other suggests that bulking is caused by a dominance of the activated sludge microbial flora by filamentous species of bacteria. The former, known as the sludge surface model, was based on the observation that the magnitude of the floc surface-charge affects: sludge settlement behaviour, thickening ability and filtration properties. The magnitude of the surface charge is governed by the exact chemical composition of the sludge surface, with the most important constituents being protein, polysaccharide, nucleic acid and lipid. Polysaccharide is likely to be the most important of these as microbial surface polysaccharides contain a monomer known as glucuronic acid. At neutral pH values this compound will carry a strong negative charge allowing the sludge to behave as a polyelectrolyte (Figure 9.9).

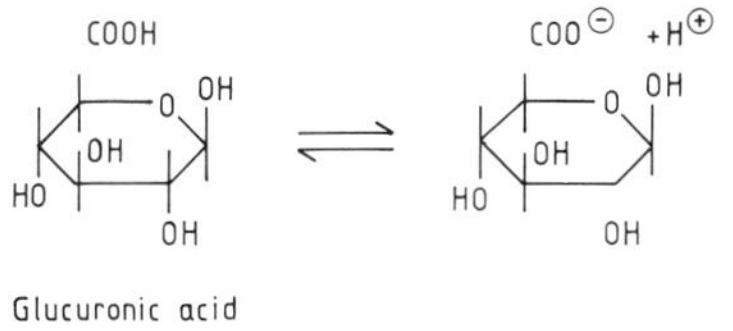

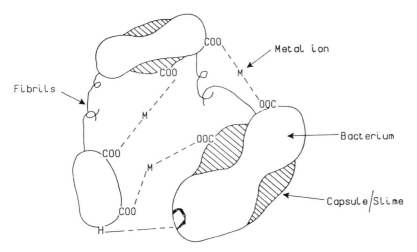

Glucuronic acid

Figure 9.9 This ionisation of the glucuronic acid molecule which occurs at pH > 3.5.

Figure 9.10 Visualisation of the interactions within an activated sludge floc, which lead to agglomeration and flocculation, as predicted by the sludge surface model. (From Forster and Dallas—Newton, 1980. Reproduced by permission of the Institution of Water and Environmetal Management.)

A scenario for flocculation between dispersed cells has been proposed whereby the production of high-molecular-weight bacterial exocellular polysaccharide bridges the distance between electrostatically neutral bacteria. This allows the end-to-end attachment of negatively charged rod-shaped bacteria into weak, shear-susceptible, long thin flocs. As more polymer is synthesised the cells become more firmly bound and the flocs stronger, allowing physical enmeshment and the binding of negatively charged sites to polyvalent cations; this type of binding is direction independent causing the flocs to become rounder. Overproduction of polymer will lead to a dispersing effect due to charge stabilisation, consequently the size and strength of a sludge floc is dependent on those factors which affect the production of exocellular polymer. Figure 9.10 is a visualisation of this model.

The second model is based on the observation that the onset of sludge bulking is almost inevitably accompanied by an increase in the number of filamentous bacteria in the sludge floc (Figure 9.11). The model suggests that there are two groups of bacteria in sludge which are responsible for flocculation: filamentous bacteria provide a rigid backbone or skeleton which gives the floc its structure and strength,

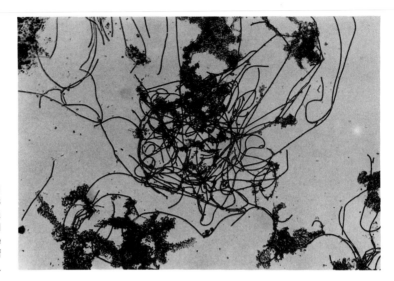

Figure 9.11 A bulking activated sludge as seen through a microscope and showing the proliferation of filamentous bacteria.

allowing colonisation by a second group of organisms, the floc-formers. These are typically *Zoogloea*-type organisms which attach to the filaments and provide a gelatinous matrix which facilitates entrapment of other microorganisms, colloidal and particulate material. Thus the settling, compaction and separation properties of an activated sludge are related to the relative numbers of filaments and floc-forming organisms. The outgrowth of excessive quantities of filaments from the floc is correlated with sludge bulking, whereas the absence of filamentous organisms leads to the production of small, weak flocs which settle poorly (pin-point flocs). An ideal sludge is

Figure 9.12 The influence of substrate concentration on the ratios of floc-forming and filamentous bacteria in an activated sludge, and its influence on the settling properties of the sludge.

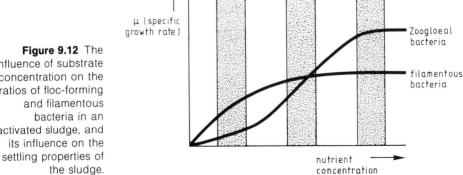

found when there is a balance between filamentous and floc-forming organisms (Figure 9.12).

The filamentous and floc-forming bacteria can be characterised by their growth and survival characteristics into three groups:

1. Floc-formers which are fast growing with a high substrate affinity, but susceptible to starvation.

2. Fast-growing, starvation-susceptible filamentous bacteria with a high affinity for oxygen.

3. Slow-growing, starvation-resistant filamentous bacteria with a high substrate affinity.

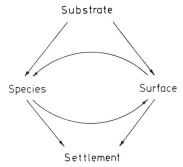

Figure 9.13 The S-hypothesis which links both the sludge surface model and the filament backbone model.

The dominant group of organisms will dictate the settling properties of the sludge and their predominance will be dependent on the relative abundance of substrate in the reactor and the prevailing dissolved oxygen concentration. Both the sludge surfaces model and the filament backbone model have much in common and their interactions have been summarised by what is known as the S-hypothesis (Figure 9.13). The filament backbone model has a major advantage, however, in that, by identification of the predominant filament type associated with a bulking episode, it offers a means of identifying the cause of the problem and suggesting corrective measures. Identification of the different types of filamentous bacteria is relatively simple and is based mainly on morphological characteristics (Figure 9.14). Twenty-two species of morphologically distinct filamentous bacteria have been identified in bulking sludges using a simple microscopic examination which can be performed in 10 min by a trained operator (Table 9.1). As more treatment plants are adopting regular microscope examination of the sludge as a diagnostic aid to plant operation, more information is being made available on the importance of each of the filament types. The importance of each type in bulking incidents have been reported for a number of countries (Table 9.2). It is now a widely held view that identification of the predominant filamentous bacteria associated with a bulking incident

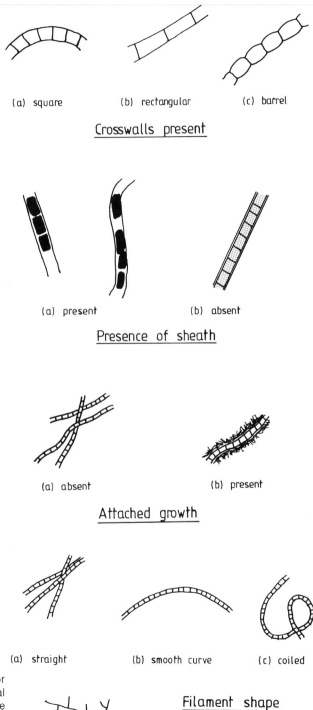

(a) square   (b) rectangular   (c) barrel

Crosswalls present

(a) present   (b) absent

Presence of sheath

(a) absent   (b) present

Attached growth

(a) straight   (b) smooth curve   (c) coiled

Figure 9.14 The major morphological characteristics of the filamentous bacteria which are used in their identification.

Filament shape

(d) mycelial

Table 9.1 Key to the identification of filamentous bacteria using morphological characteristics. (From Chambers and Tomlinson, 1982. Reproduced by permission of Ellis Horwood Ltd.)

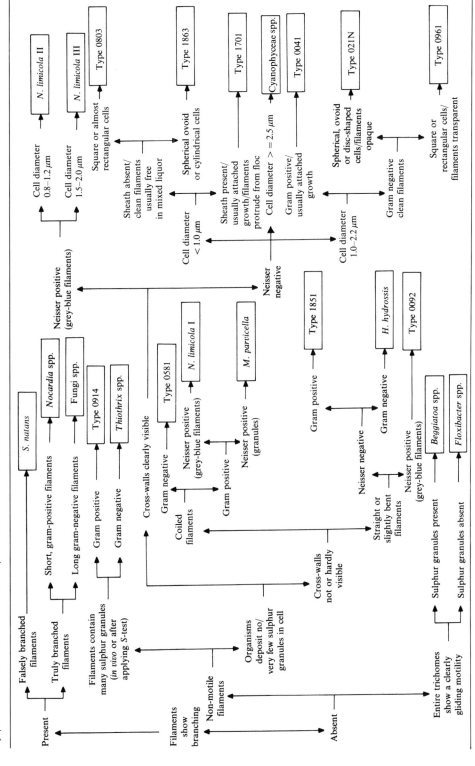

Filamentous organism	Rank			
	USA	Netherlands	South Africa	Germany
Nocardia spp.	1	14	6	5
Type 1701	2	5	8	8
Type 021N	3	2	—	1
Type 0041	4	6	6	3
Thiothrix spp.	5	19	9	—
Sphaerotilus natans	6	7	—	4
Microthrix parvicella	7	1	2	2
Type 0092	8	4	1	—
Haliscomenobacter hydrossis	9	3	9	6
Type 0675	10	—	4	—
Type 0803	11	9	7	10
Nostocoida limicola	12	11	8	7
Type 1851	13	12	3	—
Type 0961	14	10	—	9
Type 0581	15	8	9	—
Beggiatoa spp.	16	18	—	—
Fungi	17	15	—	—
Type 0914	18	—	5	—

Table 9.2 The most predominant filament types associated with sludge bulking in a number of countries.

Causative condition	Indicative filament type
Low dissolved oxygen	Type 1701, *S. natans*, *H. hydrossis*
Low F/M	*M. parvicella*, *H. hydrossis*, *Nocardia* sp., types 021N, 0041, 0675, 0092, 0581, 0961, 0803
Septic wastewater	*Thiothrix* sp., *Beggiatoa*, type 021N
Nutrient deficiency	*Thiothrix* sp., *S. natans*, type 021N
Low pH	Fungi

Table 9.3 Conditions causing sludge bulking and the associated dominant filamentous bacteria.

will provide information as to the operating conditions which have led to its proliferation. A tentative scheme for diagnosing causes of sludge bulking has been proposed already (Table 9.3) and future research should resolve many of the ambiguities in this table.

There are still many people who feel that it is unnecessary to understand the mechanisms of sludge bulking, or to perform microscopic sludge investigation. They claim that for the majority of bulking incidents, correct and efficient operation of sedimentation tanks will minimise the problem (secondary sedimentation tanks are often conservatively designed and rarely operate at design load). If this strategy proves unsuccessful, it is proposed that an 'action flow sheet' will facilitate identification of the causes of bulking (Table 9.4). Although use of such a table would eventually pin-point the reason

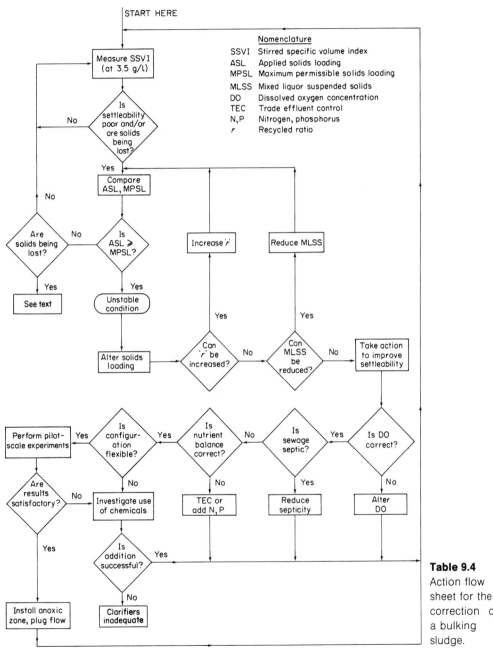

Table 9.4
Action flow
sheet for the
correction of
a bulking
sludge.

for sludge bulking, this could well prove a lengthy investigation compared to sludge microscopy, which is a potentially fast and accurate diagnostic technique.

Control of activated sludge bulking

Using either the action flow sheet or a microscopic sludge investigation (or both), it is generally possible to establish the reasons for the rapid proliferation of filamentous bacteria, and thus take appropriate remedial measures. The most common causes of bulking sludges are given under the headings below.

Low dissolved oxygen bulking

Routine measurement of dissolved oxygen should be part of standard plant operating procedure and most larger plants employ automatic control. There is some doubt about the concentration of dissolved oxygen necessary to avoid poor settleability. For instance oxygen determinations made in the mixed liquor give no indication of the O_2 gradient across a floc. An adequate mixed liquor dissolved oxygen concentration is of the order of 1 mg/l for a non-nitrifying plant and 2 mg/l for a nitrifying plant. Filamentous bacteria associated with low dissolved oxygen bulking are type 1701 and S. natans, the former suggesting a severe dissolved oxygen limitation and the latter a modest limitation.

Low organic loading

No universal relationship between loading and settling characteristics has yet been identified. During severe bulking a continued loss of solids from the reactor increases the loading rate and exacerbates problems caused by high loading rate. Changes in sludge loading also produce changes in sludge age, with an increased loading giving rise to a younger sludge as a result of increased microbial growth.

To prevent bulking problems which are associated with organic loading rates, a plant should incorporate an area of high floc loading and an area of low substrate concentration external to the flocs. This will promote the preferential growth of floc-forming bacteria, which have the ability to absorb and store organic material at high concentrations more effectively than the filamentous bacteria. The period of low substrate concentration is to allow the completion of metabolism which will regenerate the absorption capacity of the floc by the time it is recycled back to the aeration tank inlet. Filamentous bacteria are thought to be more efficient at obtaining nutrients in low concentrations as they protrude from the floc into the bulk liquor. This gives them a greater effective surface area for absorption of

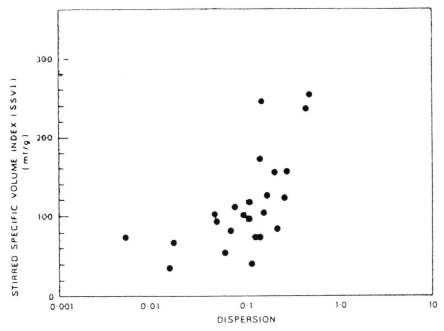

Figure 9.15 The effects of the reactor mixing regime on the settling properties of the sludge. (From Chambers and Tomlinson, 1982. Reproduced by permission of Ellis Horwood Ltd.)

nutrients and oxygen. Conditions of high substrate concentration, followed by a period of low concentration, are found in reactors with mixing regimes which approximated to plug- flow, and these do not seem to suffer bulking as much as completely mixed reactors (Figure 9.15). At small sewage works which suffer periodic bulking problems, a zone of high floc loading can be created by returning the thickened activated sludge to the outlet channel of the primary sedimentation tank. At works where this simple strategy has been adopted, dramatic improvements in sludge-settlement properties have been observed.

Nutrient-deficient wastes

If sewage is to support the growth of microorganisms it must possess certain essential nutrients in a form available for assimilation by these organisms. The ratios of the major nutrients should also be within a certain range. Domestic sewage is generally carbon limited for bacterial growth, which is an ideal situation as it means that the major polluting component is the first to be exhausted. Many industrial wastes, for example, brewery waste, paper-mill waste and fruit-processing waste, are very high in carbon, and when they form a large component of the influent, then other nutrients such as nitrogen or phosphorus may become limiting for growth. Under these circumstances certain filamentous bacteria may proliferate which do not require nitrogen and phosphorus in such large concentrations as the floc-formers. An ideal nutrient ratio of BOD:phosphorus:nitrogen is 100:1:5. Correction is easy and involves dosing the influent with the depleted nutrient in order to bring the concentration up to the required ratio. Presence of the filamentous bacteria types 021N, 0041, *H. hydrossis* and *S. natans* is associated with nutrient deficiency.

Control of bulking through chlorination

Chlorination is a last resort in the control of bulking, it should only be contemplated when effluent from the bulking plant is likely to cause environmental damage to the receiving water. If used correctly it does provide a rapid alleviation to the problems of bulking (control through changes in operating conditions frequently requires three cell residence times to alleviate the problem); if used incorrectly, however, it may completely inhibit all treatment.

Chlorine should be added to a waste at a point where the concentrations of organic material and ammonia are at a minimum in order to minimise chlorine demand. The return activated sludge line is the obvious choice but its use may be precluded owing to inaccessibility. Also at plants operated with long aeration periods, chlorine dosing at this point would not provide a sufficient period of exposure. The most common dosing points are (Figure 9.16):

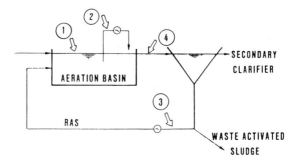

Figure 9.16 Recommended dosing points for chlorine addition in order to alleviate severe incidents of sludge bulking. The dosing points (1–4) are referred to in the text.

1. Directly into the aeration basin,

2. Into a side-stream into which mixed liquor is pumped from and returned to the aeration basin;

3. Directly to the return activated sludge line;

4. Into the mixed liquor stream between the aeration basin and the secondary clarifier.

To ensure that successful control of bulking is achieved with minimal inhibition of carbonaceous oxidation and minimum chlorine discharged to the watercourse, the following criteria must be rigidly adhered to:

1. Establish a target value for SSVI. The nomograph (Figure 9.8) may be used to calculate the maximum SSVI at which a plant will operate successfully. Chlorination should only be initiated when this value is consistently exceeded. Trend plots help in anticipating SSVI changes and in adjusting chlorine dose.

2. Add chlorine in known and controlled doses to the activated sludge at a point of excellent mixing.

3. The chlorine dose should be measured on the basis of a sludge inventory which should include sludge held in the secondary clarifier. Determination of chlorine dose is not very well defined owing to the numerous factors which might influence it. For instance filamentous bacterial species show different susceptibilities to chlorine, with those possessing a sheath being particularly sensitive. In addition different wastewaters show variations in the breakpoint chlorine demand. Doses should be determined for a given plant, although the following guidelines have been recommended: overall mass dose (g Cl_2/kg MLSS d): 1–12; concentration (mg Cl_2/l) 1–20; local mass dose (g Cl_2/kg MLSS) 2–10.

4. Because a toxicant is being added to a biological system, reliable sludge-settling and effluent-quality measurements must be made to assess its effect. Sludge settling is assessed by means of the SSVI and sludge blanket depth. Effluent quality is monitored by turbidity, and a rapid increase indicates a chlorine overdose. If chlorine is applied as a shock load the effluent frequently turns milky as a result of severe bacterial lysis.

9.4 THE USE OF PROTOZOA AS INDICATORS OF ACTIVATED SLUDGE PLANT PERFORMANCE

Growth of protozoa in wastewater treatment plants

Protozoa demonstrate a wide range of feeding modes and are capable of feeding on soluble and particulate organic material, as well as bacteria and other protozoa. They are an integral part of the food chain of wastewater treatment plants. Wastewater provides a source of organic material in the form of soluble and insoluble BOD, and any microorganism capable of assimilating organic material will grow and reproduce. These organisms now provide a food source for holozoic protozoa and a food chain has been initiated. When a wastewater treatment plant operates at a steady state, the food chain also establishes itself at a steady state, with the numbers of each species relatively stable. If there are any perturbations to the system, such as a sudden increase or decrease in the influent BOD concentration, or sudden change in sludge age caused by excessive solids wasting, the food chain will adjust itself to the new conditions. The changes in microbial composition of a reactor over a slowly changing range of environmental conditions is known as a relative predominance diagram, and a typical relative predominance diagram is illustrated in Figure 9.17.

The increase or decrease of a particular protozoal species can often be explained by examining the feeding patterns of the protozoa, together with the availability of an appropriate food source. As wastewater is introduced into a reactor, the BOD is high, the bacteria at their lowest number and therefore saprozoic protozoa such as the Sarcodina might be observed. These protozoa are inefficient at competing for food and are only observed during start-up or after recovery from toxic shock. As the numbers of bacteria increase, the Sarcodina are replaced by flagellated protozoa which are highly

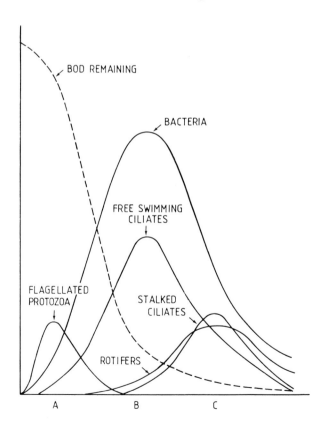

Figure 9.17 A relative predominance diagram illustrating the changes in protozoal ecology of an activated sludge as the sludge age increases and the loading rate decreases.

Relative predominance diagram

mobile and able to compete for the available food supply more effectively. Because these organisms are so active, they require a large amount of energy to maintain a stable population. They are characteristic, therefore, of a high-rate activated sludge reactor, where food is abundant (point A on Fig. 9.17). Because of their motility, they do not settle well, and settled sludges from the high-rate process are characteristically turbid. Free-swimming ciliates are also efficient feeders and are capable of surviving with a lower food supply. Consequently they soon replace the flagellates and point B on Figure 9.17 represents a conventional activated sludge plant with large numbers of free-swimming ciliates, maximum numbers of bacteria and very little residual BOD. After point B, the reactor tends towards an extended aeration system. As there is little available BOD remaining, bacteria became the predominant protozoal food supply. Stalked ciliates and the rotifers, which consume whole bacteria, now start to predominate.

The relative predominance diagram (Figure 9.17) allows us at a glance to obtain an indication of the loading rate and sludge age of a reactor simply by recognising three main groups of protozoa. Identification of protozoa to genus and species level will provide a lot more information, but requires a period of training before it can be performed with confidence. The most important protozoal species, together with their habit and food source, are given in Table 9.5. An index relating effluent quality (as BOD) has been developed based on a large programme for monitoring protozoal distribution at a number of activated sludge plants. Plant loading was shown to affect both the total number of protozoal species present and their distribution. At low loading rates the holotrich, peritrich and hypotrich population were equally distributed, and as the loading increased the total number of ciliates decreased with only peritrichs and holotrichs abundant. At very high loadings flagellates were

Protozoa	Habit	Food
Trachelophyllum pusillum	F/C	Ca
Opercularia microdiscum	A	B
Carchesium polypinum	A	B
Vorticella convallaria	A	B
Chilodenella cucullulus	F	B
Opercularia coarctata	A	B
Vorticella microstoma	A	B
V. fromenteli	A	B
Aspidisca costata	C	B
Podophyra fixa	A	Ca

A—attached growth, C—crawling forms, F/C—free-swimming and crawling B—bacteriavore, Ca—carnivore.

Table 9.5 The more important protozoa in sewage treatment processes and their characteristics.

Species	Association ratings for effluent BOD range			
	0–10	11–20	21–30	> 30
Trachelophyllum pusillum	3	3	3	1
Litonotus fasciola	0	10	0	0
L. carinatus	10	0	0	0
Hemiophrys fusidens	3	4	3	0
Chilodonella cucullulus	4	4	1	1
Colpidium colpoda	0	0	4	6
Tetrahymena pyriformis	1	3	3	3
Paramecium caudatum	2	5	3	0
Aspidisca costata	3	3	2	2
A. lynceus	5	5	0	0
Euplotes moebiusi	3	3	3	1
Vorticella convallaria	3	4	2	1
V. microstoma	2	4	2	2
V. alba	3	3	3	1
V. fromenteli	5	4	1	0
V. striata var. *octava*	3	3	2	2
V. aequilata	2	2	3	3
V. nebulifera var. *similis*	5	5	0	0
V. campanula	8	2	0	0
V. elongata	10	0	0	0
V. communis	10	0	0	0
Carchesium polypinum	3	5	2	0
Opercularia coarctata	2	2	4	2
Epistylis rotans	10	0	0	0
E. plicatilis	0	4	4	2
Spirostomum teres	0	10	0	0
Acineta grandis	10	0	0	0
Podophrya fixa	0	2	7	1
Flagellated protozoa	0	0	4	6

Table 9.6 Association ratings of protozoa found in activated sludge.

present. Based on this work, a species association was derived to predict effluent quality in the range 0–10, 11–20, 21–30 and > 30 mg BOD/l (Table 9.6). This correlation was accurate in predicting the monthly effluent BODs at 85% of the plants where it was tested.

Regular microscopic examination of sludges to genus level can also provide a wealth of information on the operating conditions in the plant. This tends to be very plant and wastewater specific and is most useful if it is performed on a daily basis. If microscopic sludge investigation is adopted as a potential aid in plant control, it is important to ensure that the operator who performs the examination is familiar with the theory of plant operation and control. The major failings in the past have been due to plant operators who were not familiar with protozoal identification receiving data from biologists who were not familiar with plant operation. Under such circum-

Observations of Protozoa	Indicator value
Peritrichs present in appreciable numbers. Crawling ciliates also present but few or no mastigophora present	Healthy, mature sludge giving a low effluent BOD. Usually nitrifying
Protozoa absent in return activated sludge (or inactive)	Too long a sludge retention time in final settlement tank
All protozoa active but Mastigophora increasing in numbers and sludge deflocculating, producing large numbers of free bacteria	Severe oxygen sag in final settlement tank due to excessive residence time, or organic shock load
Absent or sparse ciliate population consisting almost exclusively of swimming and crawling ciliates. Mastigophora present in noticeable numbers	Too high a sludge surplus rate leading to a young sludge with a high oxygen demand (hence dissolved oxygen concentration lower than expected).
Peritrichs inactive but crawling and swimming ciliates mostly active	Mild toxic shock load (possible increase in dissolved oxygen concentration due to reduced metabolic rate of bacteria). Possible loss or reduction of nitrification ability
All protozoa inactive/absent except Mastigophora, whose numbers rise appreciably.	Severe toxic shock load. Definite loss of nitrification. (larger dissolved oxygen concentration increase). High effluent suspended solids
Proportion of swimming ciliates: peritrichs increasing accompanied by a gradual increase in Mastigophora	Chronic F/M ratio leading initially to loss of nitrification followed by eventual loss of treatment

Table 9.7 The potential of protozoa to act as indicators of plant performance.

stances a lot of biological data accumulates which cannot be interpreted and this results in a potentially useful technique being viewed unfavourably. Table 9.7 indicates the sort of useful information which can result when the technique is used properly.

10 Waste Stabilisation Ponds and Aerated Lagoons

10.1 INTRODUCTION

The wastewater treatment systems discussed in previous chapters have been aerobic treatment systems designed to speed up the process of natural aeration and bio-oxidation of organic material, by forced aeration. Such systems are commonly referred to as conventional treatment and require a constant input of mechanical energy together with regular maintenance if they are to achieve consistent BOD and nutrient removal. In many parts of the world it is not possible to provide either the energy or the maintenance requirements. In addition, in these areas the main requirement of a wastewater treatment process is not the removal of BOD but the removal of excreted pathogenic microorganisms which pose the threat of infection from a wide range of water-related diseases.

Under these circumstances a wastewater treatment system, known as a waste stabilisation pond, provides a cheap alternative to conventional processes. A waste stabilisation pond is a shallow excavation which receives a continuous flow of wastewater and a

Figure 10.1 Aerial view of the Dandora pond system in Nairobi, Kenya. The photograph shows two series of a facultative and three maturation ponds. The facultative ponds each measure $700 \times 300\,m^2$ and the maturation ponds are $300 \times 300\,m^2$, giving a total area of 96 ha. The ponds were designed to treat a flow of $30\,000\,m^3$/day with a design temperature of $17\,°C$. (Photograph courtesy of D.D. Mara.)

number of these are required, generally arranged in series such that successive ponds receive their flow from the previous pond (Figure 10.1). The degree of treatment achieved is a function of the number of ponds in the series, and the retention time of the wastewater in each pond (Figure 10.2). A pond requires only simple maintenance and relies on sunlight as its only source of energy. Because ponds rely on sunlight for energy, treatment is slow, and the retention time of a pond series is measured in weeks rather than hours. This necessitates a large land requirement, which is the major drawback to the widespread adoption of ponds as a treatment option. However, in those parts of the world for which they are most suited, land is generally freely available at reasonable cost.

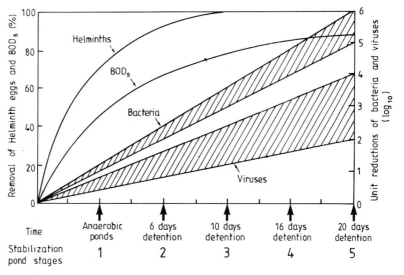

Figure 10.2 Generalised curve for the removal of organic material and excreted pathogens across a waste stabilisation series, for temperatures in excess of 20 °C. (From *IRCWD News*, **23**, 1985. Reproduced by permission of the International Reference Centre for Waste Disposal.)

Although the most important role of waste stabilisation ponds is in the removal of pathogenic microorganisms, they are still capable of producing an effluent with a low BOD and nutrient concentration. The oxygen necessary to satisfy the carbonaceous and nitrogenous oxygen requirements is provided by algal photosynthesis, and the rate at which this oxygen is produced will determine the rate at which aerobic, heterotrophic oxidation takes place; in the absence of algal-generated oxygen, BOD removal will still take place as a result of sedimentation and anaerobic metabolism. An additional advantage of ponds is that they have a very high resistance to hydraulic and organic shock loadings and can tolerate influent heavy metal concentrations up to 30 mg/l.

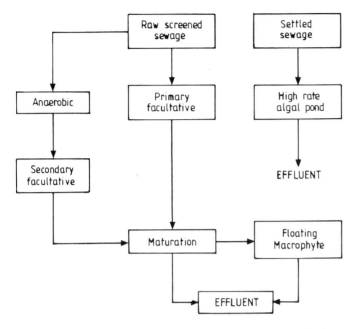

Figure 10.3 A simple classification of the major pond types based on the influents they receive.

A typical classification of the many different pond types is shown in Figure 10.3, and the major function of each of them is described in Table 10.1. The most common arrangement of ponds is to have an anaerobic pond together with a facultative pond arranged in series and followed by a number of maturation ponds, although other arrangements are possible (Figure 10.4). The number and size of the maturation ponds will depend on the requirements for the final effluent quality. The anaerobic pond is frequently omitted as it is associated with the production of odours; however, this is due more to bad design that any intrinsic features of anaerobic ponds and, in view of their efficiency at BOD removal, inclusion is recommended.

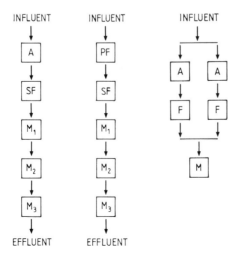

Figure 10.4 Examples of the many different arrangements of ponds in a series.

Pond type	Depth (m)	Retention time (d)	Major role	Typical removal efficiencies
Anaerobic	2–5	3–5	Sedimentation of solids, BOD removal, stabilization of influent, removal of helminths	BOD 40–60%, SS 50–70%, faecal coliforms 1 log, helminths 70%
Facultative	1–2	4–6	BOD removal	BOD 50–70%, SS increases due to algae, faecal coliforms 1 log
Maturation (for three ponds)	1–2	12–18	Pathogen removal, nutrient removal	BOD 30–60%, SS 20–40%, faecal coliforms 4 log, nitrogen 40–60%, helminths 100%

Table 10.1 The principal functions of the main pond types and their typical performance and operating data.

Because they are an entirely natural process, a complex biological ecosystem develops within the pond. Consequently, despite the fact that they are the simplest form of wastewater treatment system, they are the most poorly understood in terms of the reactions which take place within them. As a result of this, models for the design of waste stabilisation ponds tend to be purely empirical.

Effluent standards

In many arid and semi-arid areas of the world, large-scale reuse of sewage effluents is necesary because of the water shortages which result from increasing populations and agricultural demand. The health risks associated with human waste reuse have been widely examined over the past 20 years and many epidemiological studies have shown demonstrable health effects from wastewater reuse. On the basis of these studies, the relative health risks associated with reuse of untreated wastewaters have been quantified in terms of the excess infection caused by different classes of pathogens (Table 10.2). This scheme has been used as a basis for the provision of firm guidelines aimed at minimising the health risks associated with reuse of wastewaters. It has resulted in the recommendation of standards for the microbiological quality of treated wastewaters which are both technically feasible to achieve and also, on the basis of the best epidemiological evidence to date, will minimise associated health risks to an acceptable level (Table 10.3). The standards are based solely on removal of intestinal nematodes and faecal coliforms, and in

Class of pathogen	Relative amount of excess frequency of infection or disease
1. Intestinal nematodes: *Ascaris, Trichuris, Ancylostoma, Necator*	High
2. Bacterial infections: bacterial diarrhoeas (e.g. cholera), typhoid	Lower
3. Viral infections: viral diarrhoeas, hepatitis A	Least
4. Trematodes and cestode infections: schistosomiasis, clonorchiasis, taeniasis	From high to nil, depending upon the particular excreta use practice and circumstances

Table 10.2 Relative health risks from use of untreated excreta and wastewater in agriculture and aquaculture. (From *IRCWD News*, **23**, 1985. Reproduced by permission of the International Reference Centre for Waste Disposal.)

Reuse process	Intestinal nematodes[b] (geometric mean no. of viable eggs per litre)	Faecal coliforms (geometric mean no. per 100 ml)
Restricted irrigation[c] Irrigation of trees, industrial crops, fodder crops, fruit trees[d] and pasture[e]	1	Not applicable[c]
Unrestricted irrigation Irrigation of edible crops, sports fields and public parks[f]	1	1000[g]

Table 10.3 Tentative microbiological quality guidelines for treated wastewater reuse in agricultural irrigation[a]. (From *IRCWD News*, **23**, 1985. Reproduced by permission of the International Reference Centre for Waste Disposal.)

[a] In specific cases, local epidemiological, cultural and hydrogeological factors should be taken into account, and these guidelines modified accordingly.
[b] *Ascaris, Trichuris* and hookworms.
[c] A minimum degree of treatment equivalent to at least a 1-day anaerobic pond followed by a 5-day facultative pond or its equivalent is required in all cases.
[d] Irrigation should cease 2 weeks before fruit is picked, and no fruit should be picked off the ground.
[e] Irrigation should cease 2 weeks before animals are allowed to graze.
[f] Local epidemiological factors may require a more stringent standard for public lawns, especially hotel lawns in tourist areas
[g] When edible crops are always consumed well cooked, this recommendation may be less stringent.

the case of the latter a 5 or 6 log removal is required. A waste stabilisation pond is the only form of wastewater treatment which can guarantee such high removal rates without resorting to disinfection of effluents. In hot climates these standards are readily achievable by a series of five ponds, each with a retention time of 5 days; this will also produce an effluent with a low BOD and nutrient concentration which is suitable for unrestricted irrigation (Figure 10.2).

10.2 ANAEROBIC PONDS

Principles

As its name implies, an anaerobic pond lacks dissolved oxygen, and the active microbial population comprises facultative and strictly anaerobic microorganisms. The organic material which is present in the influent is therefore degraded by fermentative pathways (Figure 10.5). The degradation of organic substrates with the production of volatile acid is known as putrefaction, and end-products such as butyric acid are extremely malodorous. If metabolism were to cease at this point, then the noxious smells produced by the pond would make them very unpopular. Fortunately a group of strictly anaerobic bacteria, known as the methanogens, are able to obtain their energy for growth by coupling the oxidation of these volatile acids to the reduction of carbon dioxide, resulting in the production of methane, which is the most reduced form of carbon:

$$CH_3COOH \rightarrow CH_4 + CO_2 \tag{10.1}$$

$$CO_2 + 4H_2 \rightarrow CH_4 + 2H_2O \qquad \Delta G = -130\,kJ/mol \tag{10.2}$$

The resulting gaseous end-products are odourless and when they escape to the atmosphere they contribute to the process of BOD removal. In common with other anaerobic oxidations the energy available to the methanogens is low, and consequently they have a low cell yield; as much as 70% of the BOD removed in an anaerobic

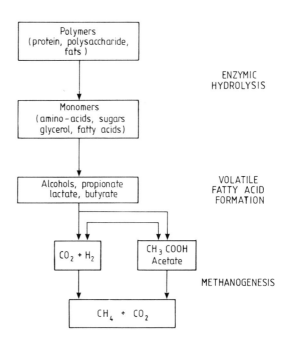

Figure 10.5 The fermentative pathways for the degradation of organic material to methane in an anaerobic pond.

pond will be in the form of methane gas. Anaerobic ponds can therefore operate for many years without desludging.

Methanogens are very susceptible to environmental conditions, in particular changes in the pH, and they will only tolerate a pH in the range 6.2–8.0. If the rate at which volatile fatty acids are produced is in excess of the rate at which they are degraded by the methanogens, then the pH will fall, the methanogens are inhibited and then ultimately killed. The growth rate of the methanogens is thus the rate-limiting step and determines the maximum organic loading to an anaerobic pond. The methanogens show a marked response to temperature, with little activity below $10\,^{\circ}C$ but increasing rapidly as the temperature increases. Temperature, therefore, is the environmental factor used in the calculation of the organic loading rate.

Because of their long retention times (up to 3 days), sedimentation provides an additional mechanism for the removal of BOD which is in the form of suspended solids, and this mechanism is independent of temperature. These sedimented solids will undergo rapid anaerobic decomposition at the bottom of the pond and, as a result of vigorous release of gaseous end-products, sludge from the pond bottom is carried to the surface. This process serves to seed the upper layers of the pond with active methanogens and also aids in pond mixing. A well-designed anaerobic pond can achieve up to a 60% reduction in BOD depending upon the temperature and retention time. As a result of the degradation of sedimented organic material, a large amount of ammonia which was previously organically bound is released as ammonium. Nitrification cannot occur owing to the lack of oxygen and this ammonium leaves the pond in the effluent. The effluent from an anaerobic pond often contains up to 20% more ammonium than was present in the influent.

Of particular importance in anaerobic ponds is the biological interconversions undergone by sulphur-containing compounds. Hydrogen sulphide evolution is responsible for the odours asociated with anaerobic ponds and it results from the anaerobic reduction of sulphate. The various forms of sulphur found in a wastewater are as follows:

Oxidation number	-2	0	$+4$	$+4$	$+6$
Formula	S^{2-}	S	SO_3^{2-}	SO_2	SO_4^{2-}
Name	Sulphide	Sulphur	Sulphite	Sulphur dioxide	Sulphate

$$(10.3)$$

The reduction of sulphate to sulphide requires eight electrons and, in anaerobic ponds, acetate is a convenient source of electrons. The reaction proceeds according to the equation:

$$CH_3COOH + SO_4^{2-} + 3H^+ \rightarrow 2CO_2 + H_2S + 2H_2O \quad (10.4)$$

This reaction is carried out by a number of sulphate reducers such as *Desulfuromonas acetoxidans* and hydrogen sulphide is the major end-product. Several organisms are capable of utilising the energy stored in this highly reduced sulphur compound by oxidising it back to sulphate. These include chemolithotrophs such as *Thiobacillus, Beggiatoa* and *Thiothrix* as well as the phototrophic green and purple sulphur bacteria such as *Chlorobium* and *Rhodospirillum*. The absence of light precludes significant activity by the phototrophs and the chemolithotrophs require micro-oxic conditions (i.e. very low oxygen concentrations). As a result of this, if the concentration of either sulphate or the electron donor is too high, then the production of odorous H_2S occurs. This is easily eliminated by ensuring that the influent sulphate concentration is below 500 mg/l and that the influent organic strength is less than 400 mg BOD/l.

Design

In order to attain anaerobic conditions in a pond the oxygenation due to algal photosynthesis must be inhibited, and the rate of oxygen utilisation within the pond must exceed the rate of reaeration at the surface. This is achieved by designing the pond as deep as the site conditions will allow (to an optimal depth of 4 m) and ensuring a high organic loading rate. Consequently the basis for anaerobic pond design is the volumetric organic loading rate (λ_v) expressed in g/m^3 d. The volumetric organic loading rate is defined as

$$\lambda_v = \frac{L_i Q}{V} \tag{10.5}$$

where L_i is the average influent BOD (mg O_2/l), Q the average influent flow rate (m^3/d) and V the required pond volume (m^3).

As anaerobic processes are very susceptible to temperature changes, the required value of λ_v is selected on the basis of the mean temperature of the coldest month. There are currently no suitable deterministic models available to calculate values of λ_v and a simple empirical technique is employed. For temperatures below 10 °C, a value for λ_v of 100 g m^3 d is selected and for temperatures above 20 °C, 300 g/m^3 d is used. Between these two temperatures the volumetric loading is calculated from the relationship:

$$\lambda_v = 20T - 100 \tag{10.6}$$

The degree of BOD removal at the chosen volumetric loading is calculated in a similar manner from the equation:

$$\text{BOD removal } (\%) = 2T + 20 \tag{10.7}$$

Knowing the volumetric loading, the required pond volume is calculated from equation (10.5)

10.3 FACULTATIVE PONDS

Principles

The major role of facultative ponds is for the removal of BOD. The presence of both aerobic and anaerobic (strictly speaking anoxic) environments within the same pond means that both anaerobic metabolism associated with anaerobic environments, and oxidative metabolism associated with a wholly aerobic environment will occur. The combination of aerobic and anaerobic metabolism within the same body of water allows complete nutrient cycling to take place, and complete cycles for carbon, nitrogen and sulphur are possible under the appropriate conditions. Unlike conventional processes, the oxygen required to satisfy the carbonaceous and nitrogenous oxygen demand is supplied by biological means and not mechanically. This is achieved by exploiting the elevated temperatures and sunlight, and providing conditions under which photoautotrophic organisms (principally algae), can utilise the carbon dioxide evolved in hetero-trophic metabolism. This is metabolised via the Calvin cycle to provide them with a source of carbon, using energy from photosyn-thesis, resulting in the production of gaseous oxygen. This relation-ship between the algae and bacteria can be considered as a form of symbiosis and their interactions have been summarised in Figure 4.13.

In order to encourage the growth of algae, the loading to a facultative pond must be controlled so that the oxygen demand of the influent wastewater does not exceed the rate at which oxygen can be supplied by photosynthesis. Because of the requirements of the algae for light, the rate of oxygen production and thus the concentration in

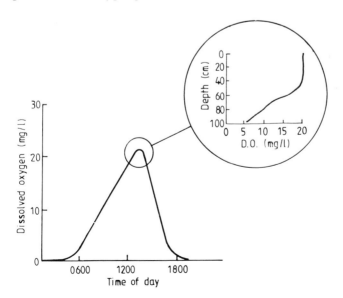

Figure 10.6 The variations of oxygen within the water column of a facultative pond, which demonstrate both diurnal and spatial variations.

the pond, will vary both diurnally and with pond depth (Figure 10.6). As the pond depth increases and the light penetration decreases, less oxygen is available. Although the oxygen supplied close to the surface is distributed throughout the pond by mixing and diffusion, the lower half of the pond is generally devoid of oxygen, and hence anaerobic metabolism occurs.

The photosynthetic activity of algae in facultative ponds varies with the intensity of the incident sunlight. In bright sunlight, the algae form dense bands in a layer up to 50 cm deep at the surface of the pond. Their rate of oxygen production is frequently so rapid that it is produced faster than it can diffuse to the atmosphere, and supersaturated oxygen concentrations are attained. In order to support this intensive rate of photosynthesis the algae utilise a large amount of carbon dioxide, and as a result the pH at the algal band can reach as high as 9.5. Below this surface layer the oxygen concentration declines rapidly as it is utilised by the heterotrophic bacteria for aerobic respiration.

During daylight the majority of the BOD is removed by facultatively aerobic bacteria at the oxic surface layer. As the intensity of the incident sunlight decreases, photosynthetic activity declines until at low light levels the algae switch from photosynthesis to respiration. During the hours of darkness, as the residual oxygen is utilised, the pond will slowly become anaerobic. Anaerobic metabolism in the pond sediment serves to degrade sedimented sludge, and thus increases the times between desludging. Typically the desludging period is from 5 to 10 years. Anaerobic metabolism in the sediment also results in the formation of gases such as nitrogen and hydrogen sulphide which carry the sludge particles to the surface. When the pond surface is aerobic the oxygen causes a rapid chemical oxidation of the hydrogen sulphide; however, when the pond is completely anaerobic, rising sludge seeds the upper layers with methanogens and this prevents putrefaction taking place.

A large amount of nitrogen removal takes place in facultative ponds, but as yet the mechanism for this is unclear. A fraction will be bound in the sediment as organic nitrogen associated with biomass and removed during desludging, but this is unlikely to account for all the nitrogen removed. The alternate aerobic/anaerobic conditions would suggest nitrification followed by denitrification as the most likely removal mechanism. This has not yet been substantiated, however, and the numbers of nitrifying bacteria isolated from facultative ponds are low. A third mechanism has been proposed in which the ammonium produced in the sludge layer by anaerobic degradation or organic material (in particular proteins) is converted to ammonia gas at the high pH's of the active algal band, according to the equation:

$$NH_4^+ + OH^- \underset{pH < 9.0}{\overset{pH > 9.0}{\rightleftharpoons}} NH_3 + H_2O \qquad (10.8)$$

The gaseous ammonia is then lost to the atmosphere, and the

process is known as volatilisation. However, ammonia is a very soluble gas, and does not pass out of solution into the gaseous phase, unless displaced by other gases. It is unlikely that volatilisation will be the major mechanism for nitrogen removal, given the relatively quiescent conditions in a pond.

The mechanism by which sulphur is removed from facultative ponds has also received little attention. This is largely because of the complexity of sulphur metabolism which is carried out by both phototrophic and chemotrophic organisms. A simplified sulphur cycle is illustrated in Figure 10.7 and, as a result of both aerobic and aerobic conditions within the pond, all these reactions may occur to a greater or lesser extent. The predominant chemical form of sulphur within facultative ponds is important, and helps to determine the pond species ecology. As a result of anaerobic metabolism in the sediment layer, sulphate is reduced to sulphide by sulphur-reducing bacteria according to equation (10.4). In the aerobic layer this sulphide can be oxidised to elemental sulphur by colourless sulphur oxidisers such as _Beggiatoa_, which store sulphur granules. Purple and green sulphur bacteria are able to oxidise sulphide to elemental sulphur under anaerobic conditions, in the presence of light. The green Chlorobacteriaceae release this sulphur into the pond, whereas the purple Chromatiaceae retain it as sulphur granules. Both of these organisms are found in large numbers in facultative ponds and are likely to be the most important sulphur oxidisers. Excessive concentrations of sulphide in ponds are detrimental to the growth of algae if present in an undissociated form (H_2S), and the proportion of undissociated sulphide increases as the pH decreases. At the pH range normally associated with facultative ponds, a total sulphide concentration of 8 mg/l will inhibit algal photosynthesis. Under these

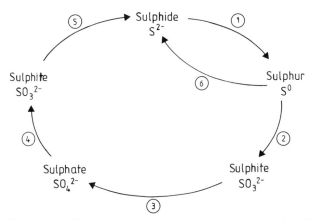

Figure 10.7 A simplified sulphur cycle which operates in a facultatively anaerobic pond. Reactions 1–6 are carried out by the following representative organisms: (1) _Beggiatoa, Thiobacilli,_ photosynthetic sulphur bacteria; (2) _Thiobacilli;_ (3) _Desulfovibrio, Thiobacilli,_ (4) sulphate-reducing bacteria (_Desulfovibrio, Desulfotomaculum_); (5) sulphate-reducing bacteria (_Desulfovibrio, Desulfotomaculum_); (6) _Desulfuromonas, Campylobacter._

conditions the Chromatiaceae will completely replace the algal population and, as they perform anoxygenic photosynthesis, the pond becomes anaerobic. When such conditions prevail it is an indication of overloading, and the loading rate must be reduced in order to re-establish an oxygenic algal population.

Design

Where a facultative pond receives waste directly without pretreatment, it is referred to as a primary facultative pond, whereas if the wastewater has previously been settled, such as the effluent from an anaerobic pond or septic tank, the pond is known as a secondary facultative pond. Primary ponds are used mainly in the treatment of weaker wastes, or in environmentally sensitive areas where the potential risk of anaerobic pond odours is unacceptable.

Early models for the removal of BOD in primary facultative ponds, assumed that they behaved as completely mixed reactors, with BOD following first-order removal kinetics. It is more common, however, to adopt an empirical approach to design based on the permissible surface loading ($\lambda_{s(max)}$ in kg/ha d). A number of such equations have been proposed based on regression analysis of data from a number of ponds world-wide. An early example of such an approach took the form:

$$\lambda_{s(max)} = 60.3(1.099)^T \qquad (10.9)$$

where T is the temperature of the coldest month in °C. This has since been modified in a linear form to incorporate a margin of safety as

$$\lambda_{s(max)} = 20T - S \qquad (10.10)$$

Values of S in the range 60–120 have been proposed.

Where ponds are designed for use in Mediterranean Europe, a more conservative form of equation (10.10) is used where

$$\lambda_{s(max)} = 10T \qquad (10.11)$$

Operating experience from existing pond installations has shown that none of the above equations prove adequate for pond design over a wide range of temperatures. An alternative approach suggests an equation incorporating a linear Arrhenius function, which takes the form:

$$\lambda_{s(max)} = a(b - c)^T \qquad (10.12)$$

By the imposition of boundary conditions for the optimum loadings at a given temperature (100 kg/ha d at 10 °C; 350 kg/ha d at 25 °C and 500 kg/ha d at 35 °C) then equation (10.12) can be written as

$$\lambda_{s(max)} = 350(1.107 - 0.002T)^{T-25} \qquad (10.13)$$

This is currently the recommended design equation for facultative ponds.

Suitable equations for the design of secondary facultative ponds take into account the fact that the active sludge layer in these ponds is not as deep as in primary ponds, due to the prior settlement which the influent has received. This will result in a reduction in the degree of BOD removal and a correction factor (0.7 is generally used) is applied to whichever primary pond equation has been selected.

The required pond area for a given areal BOD loading (calculated as the mid-depth area, A_f), is found from the equation:

$$A_f = \frac{10 L_i Q}{\lambda_s} \qquad (10.14)$$

The mean hydraulic retention time (t_f) of a facultative pond can then be determined from equation (10.14) since:

$$t_f = \frac{A_f D_f}{Q} = \frac{10 L_i D_f}{\lambda_s} \qquad (10.15)$$

where D_f is the mean depth of the facultative pond in metres. This is typically around 1.5 m and retention times of about 20–40 days are normally expected (Table 10.2). Under these conditions an effluent BOD in the range 50–70 mg/l is achieved.

10.4 MATURATION PONDS

Principles

The major role of maturation ponds is in the removal of pathogenic microorganisms such as the viruses, bacteria and helminths. This is achieved by providing a retention time long enough to reduce the numbers to the required level. The number and size of maturation ponds will therefore depend on the standard of effluent which is required; this is normally expressed as faecal coliforms/100 ml. Rational design of maturation ponds is hindered by a lack of information on the mechanism or mechanisms of pathogen removal. There is general agreement that the excreted eggs of helminths such as _Ascaris, Trichuris_ and _Taenia_ are removed by sedimentation, due to their large size (from 20 to 70 μm). Protozoal cysts of organisms such as _Giardia_, and _Entamoeba_ behave in a similar way, although they are generally much smaller (14 μm long and 8 μm broad in the case of _Giardia_), and thus require longer retention times. Removal will take place across the pond series and complete removal can be expected in ponds with overall retention times of 11 days or more. Although they are removed from the pond effluent, they are not necessarily inactivated, and can remain viable in the sludge layer for several years. This is an important consideration during desludging.

Over 100 different viruses are excreted in faeces by man, and very little is known about their survival in ponds. As most viruses carry a strong negative charge, it is assumed that the major mechanism for their removal is adsorption to particulate material, followed by sedimentation. The limited data available suggest that ponds with overall retention times of greater than 30 days, should achieve at least a 4 log reduction in enteroviruses and a 3 log reduction in numbers of rotaviruses. In a similar way to the helminths, removal from the effluent does not mean inactivation, and viruses may be capable of remaining viable in the sludge layer for long periods.

The faecal coliform bacteria have been universally adopted as indicators of excreted pathogen removal, consequently much of the work into elucidation of removal mechanisms has been carried out using these bacteria. The rate of die-off of faecal coliforms in maturation ponds is much faster than in other pond types and increases with increasing temperature. A number of mechanisms have been suggested for pathogen removal in maturation ponds of which the most important ones are nutrient starvation, enhanced pH, high dissolved oxygen concentration, lethal UV irradiation from sunlight and protozoal predation. All of the above mechanisms are closely interlinked; for instance increased sunlight causes increased UV irradiation, and by stimulating algal photosynthesis it leads to an increased pH and dissolved oxygen concentration. It is difficult, therefore, to delineate and quantify the contribution of each potential removal mechanism, thus this approach has not yet featured in pond design.

In addition to reduced pathogens, the effluent from maturation ponds is also lower in suspended solids. The suspended solids content of a facultative pond effluent is composed mainly of algae. These are predominantly motile flagellates which form dense bands to most effectively utilise the incident sunlight. In maturation ponds these are replaced by non-motile algae which outgrow the flagellates due to the increased light penetration. As these algae are non-motile, they are distributed evenly throughout the pond and their concentration in the effluent is reduced.

As well as their role in pathogen removal by pH elevation, algae play a further role in maturation ponds by removing nitrogen and phosphorus. The major mechanism of phosphorus removal is sedimentation as organic phosphate, associated with the algal cell. In addition, at elevated pH values, phosphate becomes insoluble and chemical precipitation occurs. Many algae, such as the Cyanophytae, are able to store phosphate in the form of granules, and algae also represent the largest fraction of bound organic phosphate in the pond. The transition from the anaerobic layer of a facultative pond to an aerobic pond theoretically provides ideal conditions for luxury uptake of phosphate by *Acinetobacter calcoaceticus*, but there are as yet no reports of this organism contributing to phosphate removal in ponds.

A similar removal mechanism appears to operate for nitrogen in maturation ponds. Soluble nitrogen is almost exclusively in the form of ammonia, and this form is taken up by algae in preference to nitrate nitrogen. This nitrogen will enter the pond sediments as organically bound nitrogen when the algae die and settle, and although a fraction of it is biodegradable, up to 60% remains undegraded in the pond sediment.

Design

The removal of faecal coliform bacteria from any pond follows a first-order removal kinetics and if complete mixing is assumed then:

$$N_e = \frac{N_i}{1 + K_b t} \qquad (10.16)$$

where N_e is the number of faecal coliforms in effluent (/100 ml), N_i the number of faecal coliforms in influent/100 ml, K_b = first-order faecal coliform removal constant (d^{-1}) and t = retention time in pond (d), and for a number of ponds (N) in series, equation (10.16) takes the form:

$$N_e = \frac{N_i}{(1 + K_b t_a)(1 + K_b t_f)(1 + K_b t_m)^n} \qquad (10.17)$$

where t_a, t_f, t_m are the retention times of the anaerobic, facultative and maturation ponds in the series and n the number of maturation ponds required. The first-order removal constant K_b is a lumped parameter which takes into account all the factors which affect pathogen removal. It is particularly sensitive to temperature and this effect has been modelled empirically as

$$K_{b(T)} = 2.6(1.9)^{T-20} \qquad (10.18)$$

As more quantitative information becomes available on the removal of pathogens, then deterministic equations for K_b can be formulated.

In order to use equation (10.17) to design maturation ponds, a value for both the number of maturation ponds (n), and their retention times (t_m) is required. Equation (10.17) is generally solved by iteratively increasing n, in order to find the combination of n and t_m which has the least areal land requirements. This occurs with a maximum value of n and consequently a minimum value of t_m, with the following boundary conditions:

1. The minimum acceptable value of t_m is 3 days, below which the danger of hydraulic short-circuiting becomes too great.

2. The value for t_m should not be higher than that of t_f.

3. The areal BOD loading on the first maturation pond does not exceed the areal BOD loading on the facultative pond, assuming a

BOD removal of 70% in both the anaerobic (if used) and the facultative pond.

The value of N_i should ideally be obtained from analysis of the wastewater for which the pond is intended to treat. If this is not available at the design stage, then a conservative value of 1×10^8 faecal coliforms/100 ml is often recommended. The value selected for N_e will depend upon the intended use of the pond effluent. In view of the high effluent quality which a pond is capable of producing, the effluent is frequently exploited for reuse, either in agriculture or aquaculture. Typical guideline values for the microbiological quality of treated pond effluents intended for agricultural reuse are given in Table 10.3. Thus if a value for N_e of 1000 faecal coliforms/100 ml is selected, then the effluent from the pond should be suitable for unrestricted irrigation, which includes irrigation of edible crops.

10.5 OTHER CONSIDERATIONS IN POND DESIGN

Mixing and short-circuiting

The mixing of pond contents is an important mechanism which helps to convey oxygen produced at the pond surface to the lower layers. In addition it contributes oxygen in its own right by the process of reaeration. This mixing also helps to reduce thermal stratification which aids in dispersing bacteria and algae throughout the pond, thus producing an effluent of a more consistent quality. In order to facilitate wind-induced mixing, the longest dimension of the pond should lie in the direction of the prevailing wind. As thermal stratification is more pronounced during the summer months, the direction of the prevailing wind during this period should be selected, if it is seasonally variable. Care must be taken in site selection, however, to ensure that ponds are sited downwind of the community, in case problems arise through pond odours. It is often recommended that an anaerobic pond be sited at least 1000 m away from the community, and facultative ponds 500 m away.

Short-circuiting in ponds is a common problem especially in the deeper anaerobic ponds. It can result in stagnant zones within a pond which can be responsible for odours, and it also reduces the pond volume available for wastewater treatment. These effects may be minimised by careful consideration of pond geometry, and also by sensible placement of inlet and outlet structures. For an anaerobic or primary facultative pond which receives unsettled sewage, it is important to avoid the formation of sludge banks near the inlet structure. Thus the design of these ponds approximates a rectangle with a length to breadth ratio of < 3, with the inlet and outlet

Figure 10.8 The use of baffles within a pond to reduce the degree of short-circuiting.

structures placed at diagonally opposite corners. Facultative and maturation ponds should be designed to approximate plug-flow mixing as far as is possible and thus high length to breadth ratios are employed. Site constraints frequently mean that the required length to breadth of up to 20:1 cannot be achieved. In such situations, baffles placed within the pond will help to ensure plug-flow conditions (Figure 10.8). As these ponds should lie parallel to the direction of the prevailing winds, badly placed inlet and outlet structures can favour transport of inlet wastewater directly to the outlet by surface wind action.

Siting of the inlet pipe such that the wastewater flows against the prevailing wind will effectively prevent wind-induced short-circuiting. In addition, elevated inlet pipes will produce better mixing and dispersion of the influent due to the turbulence they create. They have the added advantage that samples of pond influent can be taken easily. In large ponds the influent is often split into a number of inlets. Outlet pipes are generally (but not always), sited at the opposite end of the pond to the inlet and should always be fitted with scum guards to prevent the discharge of accumulated scum into the next pond. The level at which the pond contents are withdrawn is controlled by the depth of the scum guard, and the height of the pond controlled by the height of the outlet pipe within the scum guard. In facultative ponds the guard should extend below the level of the algal band (usually no more than 60 cm below the surface), to ensure that the effluent does not have an excessive algal content. It is often convenient to fit outlet devices with facilities for varying the level of the pond, to permit essential maintenance such as desludging and repairs to the base and embankment. A number of simple inlet and outlet devices are illustrated in Figure 10.9.

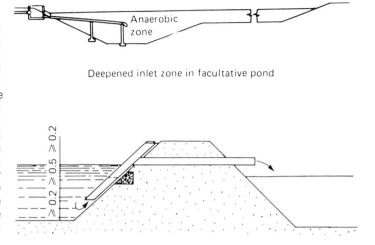

Figure 10.9 Simple inlet and outlet structures which can be used in waste stabilisation ponds: (a) the inlet to a facultative pond showing a deepened inlet zone; (b) pond outlet structure. (From Pescod and Arar, 1988. Reproduced by permission of the Food and Agriculture Organisation of the United Nations.)

Inlet works

There are a number of arguments for the against the provision of facilities for removal of coarse and floating materials and grit from a wastewater, prior to it entering a pond. The provision of screens at the inlet requires regular maintenance to clean the screens and dispose of the screenings, and at small ponds this is not available. In addition the material which is removed as screenings would either settle to the bottom of the pond, or float and collect in the corners, where it can be removed during periodic maintenance. At larger ponds, which have at least one permanent operator on site, then provision of screens is advisable. When they are provided, it is important to ensure that they are raked regularly (in the case of manually raked screens), and that adequate provisions are made for the disposal of the screenings. At smaller works it is usual to provide a single screen of 50 mm spacing to prevent very large floating solids entering the pond.

Facilities for grit removal are only required if the pond receives sewage from a combined sewerage system when the influent frequently contains appreciable quantities of grit and sand, which is associated with stormwater. It can rapidly silt up an anaerobic pond, reducing significantly the period between desludging. The most appropriate type of grit chamber is a constant-velocity channel, generally with a trapezoidal cross-section. These are manually degritted at intervals, and the grit disposed of, bearing in mind that it will be associated with both excreted pathogens and putrescible organic material. For separate sewerage systems the amount of grit and slit is negligible, and accounts for as little as 5% of the weight of the sedimented material in an anaerobic pond.

Some form of flow metering is essential at any pond installation, regardless of how small it is. Without some indication of the wastewater flow entering the pond, it is impossible to determine the loading on the pond. This is important in the case of underloaded ponds to estimate by what factor this loading could be increased or, alternatively, by what area the existing ponds need extending when overloading is experienced. A large number of simple flow-measuring devices are available such as the V-notch weir and the Venturi flume. Where flowmeters are fitted to both the influent and effluent, they provide information on rates of evaporation and infiltration and can also show when leakages are occurring within the pond. It is easy to neglect flowmeters and they are usually the first thing to fail at pond installations. If they are to serve any useful function, regular maintenance and calibration is essential.

10.6 SPECIALISED POND TYPES

Macrophyte ponds

A macrophyte is an aquatic plant which can grow either by floating on the pond surface with large suspended root systems—in which case it is termed a floating macrophyte—or it is attached by roots to the bottom of the pond and grows submerged or emergent, when it is known as a rooted macrophyte. Examples of floating macrophytes

Figure 10.10 The floating macrophyte *Eichhornia crassipes* (water hyacinth), growing in Northeast Brazil.

include *Eichhornia, Pistia* and *Salvinia*, whereas the most widely used emergent species are *Phragmites* and *Scirpus*. The most commonly used species is *Eichhornia* or water hyacinth which a floating macrophyte that can double in mass every 6 d in tropical conditions (Figure 10.10). The macrophyte biomass is harvestable and production rates of up to 250 kg dry weight/ha d of *Eichhornia* can be produced. The harvested macrophytes have many potential uses such as poultry feed supplement, cattle fodder or as a substrate for anaerobic digestion resulting in the production of methane.

Macrophyte ponds can be considered as polishing ponds which treat the effluent from a pond series. They are intended to remove the algae which are always present in maturation pond effluents, and as a result of this reduce both the suspended solid and BOD content of the final effluent. In addition the macrophytes remove nitrogen and phosphorus during growth, and these nutrients are removed from the pond when the macrophytes are harvested. It has also been suggested that the extensive root systems of the floating macrophytes will provide an ideal surface for attachment of nitrifying bacteria, and thus encourage nitrification. Algal removal is achieved because the macrophytes form a dense mat over the surface of the pond and thus prevent light penetration. Rooted macrophytes also provide an ideal environment for a microinvertebrate known as a *Daphnia* or water-flea, to breed. *Daphnia* is a natural predator of algae and can remove large quantities through grazing. If macrophyte ponds are to be considered in a pond series, it is important to ensure that the required faecal coliform removal has been achieved previously. The removal of algae by these ponds means that neither a high oxygen concentration nor a zone of high pH is produced; in addition penetration by UV light is reduced, consequently faecal coliform removal in these ponds in negligible.

Although macrophyte ponds appear to have many potential uses, there are problems associated with them. Regular cutting of rooted macrophytes is required in order to prevent decaying vegetation from silting up the pond. Floating macrophytes with small root systems show little adherence to the pond surface and are easily blown away in the wind, or collect at one end of the pond. Floating macrophytes also provide breeding-grounds for flies and mosquitoes. The *Culex* mosquito is capable of breeding in polluted waters, providing that vegetation is available. In many parts of South America, East Africa and India this mosquito is associated with the transmission of filariasis and provision of macrophyte ponds may inadvertently provide an additional breeding-ground.

High-rate algal ponds (HRAP)

A HRAP is designed primarily as a unit for carbon conversion by providing conditions for optimal algal productivity and not optimal purification efficiency. The algae thus produced are harvested in order to offset the operational costs. In order to ensure maximum light penetration, these ponds are shallow (20–45 cm) with rentention times of 1–3 days. They are typically arranged in an 'endless channel' configuration in a similar way to oxidation ditches. The bottom of the pond may need lining with impervious material and the ponds are mixed regularly to prevent the formation of a sludge layer. Typical loadings of 350 kg/ha d can yield > 100 000 kg dry weight algae/ha yr, with a protein content of 40–50%. When operated under such high loading rates the formation of algal monocultures is encouraged and typical species include *Scenedesmus*, *Euglena* and *Chlorella*.

The major aim of HRAPs is to produce and harvest algae, but to achieve the latter economically has proved very difficult due to the small size of the algae ($< 20 \mu$m). A number of standard solid/liquid separation techniques have been tried including flotation, centrifugation, flocculation with lime or alum and microstraining. These have met with limited success. The resulting algal slurry will generally have a dry solids content of 1–2% and dewatering is required to increase the solids content to 15%. In this form it is suitable for use as animal feed or for anaerobic digestion to methane gas. In order to produce a more saleable product, heat drying or vacuum filtration is necessary to increase the solids content to 90% or more. If algal removal is very efficient it is possible to achieve an effluent BOD of 25 mg/l, although faecal coliform removal is low. HRAPs have moved far away from the original concept of waste stabilisation ponds as simple low-cost treatment systems, and they find only limited use in a number of specialist applications.

10.7 AERATED LAGOONS

Principles

An aerated lagoon resembles a waste stabilisation pond in that it is a shallow basin between 2 and 5 m deep, with a large surface area, which receives a continuous flow of wastewater. It differs, however, in that the oxygen for BOD removal is provided not by algal photosynthesis but by mechanical aeration. In most temperate climates the reduced levels of sunlight mean that algal photosynthesis is also much reduced, and thus longer retention times (and therefore a larger land requirement), are required in order to ensure acceptable BOD removal. It was thought that this land requirement could be reduced by providing supplementary oxygen using mechanical aerators. A similar philosophy is often followed for the up-rating of overloaded facultative ponds in hot climates. In ponds where mechanical aerators have been installed, there is a drastic change in the ecology of the lagoon, leading to a complete disappearance of the algae and their replacement by a mixed heterotrophic bacterial community, which grows in the form of flocs which resemble activated sludge flocs. Thus, ecologically, aerated lagoons most resemble an activated sludge process operated without cell recycle. It is apparent, therefore, that instead of supplying supplementary oxygen, aerators must supply the total oxygen demand of the wastewater. In addition, as a result of floc formation, the microorganisms will agglomerate and settle under quiescent conditions; aerators must also provide adequate energy for mixing, which keeps the solids in suspension and disperses the dissolved oxygen throughout the basin.

Power inputs to aerated lagoons are not as high as those for activated sludge and larger solid particles are able to settle out. When lagoons are operated in climates where the bottom sludge temperature exceeds 20 °C, this sludge is removed by digestion. If the temperature falls below this, then solids accumulate faster than they can be digested, with a resultant sludge build-up. Under certain climatic conditions, the digestion rate during the summer months is adequate to reduce sludge accumulated during winter.

Design

The microbial community of an aerated lagoon resembles that of an activated sludge reactor in which BOD removal follows first-order removal kinetics. The large area and shallow depth of lagoons, together with the high degree of mixing provided by the aerators, means that their flow regime closely approximates a completely mixed reactor, and thus they can be modelled using equation (3.31). The first-order BOD removal constant is known to be temperature dependent and this dependence takes the form:

$$k_T = k_{20}(1.035)^{T-20} \qquad (10.19)$$

Values for k are very waste specific, but for domestic wastewaters values of k determined at 20 °C are in the range 4–6.5 d^{-1}. Design is based on the minimum winter temperature of the influent wastewater, and in hot climates aerated lagoons typically operate at hydraulic retention times in the range 2–10 days. In temperate climates the retention time can be as long as 100 days.

Assuming that the wastewater to be treated is adequately characterised, then an alternative design approach is to use the Monod equation (6.8) to calculate the solids retention time necessary to produce a given effluent quality. In the absence of cell recycle the solids retention time will be identical to the hydraulic retention and equation (6.8) can be written in the form:

$$\frac{V}{Q} = \frac{K_s + S}{\mu_m S} - \frac{1}{k_d} \qquad (10.20)$$

Again temperatures has a profound effect on the rate of BOD

Parameter	Quebec (Canada)	Lagoon Netanya (Israel)
Influent BOD (mg/l)	150	300
Effluent BOD (mg/l)	20	110
Influent suspended solids (mg/l)	100	320
Effluent suspended solids (mg/l)	20	220
Volume (m³)	4150	14 000
Depth (m)	4	4
Hydraulic retention time (d)	7	5.2
Temperature (°C)	10	28

Table 10.4 Performance and operating conditions of aerated lagoons in a range of climates.

removal, although many people have observed that above 20 °C the BOD removal rate does not seem to vary significantly. Below this temperature the coefficients k_d and μ_m are corrected using Arrhenius functions of the form:

$$\mu_m = \mu_{m(20)}(1.1)^{T-20} \tag{10.21}$$

$$k_d = k_{d(20)}(1.05)^{T-20} \tag{10.22}$$

Typical operating and performance data for aerated lagoons in a range of climates is given in Table 10.4 and whereas good BOD removal is achieved in hot climates, temperate conditions regularly result in the effluent quality exceeding 40 mg/l. In addition there is a negligible removal of pathogenic microorganisms.

Aerators

Aerated lagoons generally employ mechanical surface aerators, which are either floating or fixed (Figure 10.11). Occasionally, in deeper lagoons, submerged turbines must be provided in order to provide adequate mixing. Concrete pads are usually placed under the mixers in order to prevent scour of the lagoon bottom. These are not required, however, when the lagoon has been completely lined. Design and selection of aerators for aerated lagoons is the same as for the activated sludge process as regards the requirements for oxygen transfer. Thus the amount of oxygen required (in kg O_2/d) is calculated from equation (7.40) and a suitable aerator selected from manufacturer's data, remembering that the oxygen transfer rate under standard conditions must be corrected for field conditions

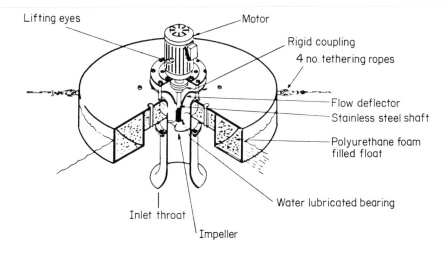

Figure 10.11 Mechanical surface aerators used in the aeration of aerated lagoons: (a) cut-view of a floating surface aerator; (b) mounted surface aerators in operation at an aerated lagoon in Northeast Brazil.

using equation (3.50). However, in addition to the aeration requirements, installed aerator power also requires a component for mixing. In order to obtain maximum removal of settleable solids, aerator power levels must be kept below a threshold value. The power level required for mixing is a function of the lagoon size, geometry and the concentration of suspended solids. A generalised relationship of the power required for mixing using a low-speed surface aerator is

$$p \ (\mathrm{W/m^3}) = 0.004X + 5 \quad \text{for values of } X < 2000 \, \mathrm{mg/l} \quad (10.23)$$

where X is the concentration of suspended solids in the lagoon. As the suspended solids concentrations in aerobic lagoons are typically in the range 200–300 mg/l, this gives a power requirement for mixing of $\sim 6\,\text{W/m}^3$. As aerator power levels are increased, there is a linear increase in the concentration of suspended solids which can be kept in suspension. If the BOD or solids loading to the lagoon increases then more aerators can be added as required, but a minimum spacing is required to prevent interference and for 75 kW aerators 20 m is required.

Effluent treatment

The effluent from an aerated lagoon requires some form of settlement stage before it is fit to discharge to a watercourse. Although this can be a conventional sedimentation tank, it is more usual to discharge into one or more ponds, dependent upon the final effluent quality required. The suspended solids content of aerated lagoon effluent is high and thus the first pond in the series acts as a settlement tank. A retention time of up to 10 days is required and depths are typically 2 m. The remaining ponds in the series behave as maturation ponds and their number and size will depend upon the degree of faecal coliform removal required. They can be designed using equation (10.17), assuming a value for N_i of 10^6 faecal coliforms/100 ml.

Where a conventional sedimentation tank is employed both the effluent and sludge will have a high pathogen content. Slow sand filters have proved successful for treating the clarified effluent, particularly in the removal of helminth eggs and protozoal cysts. The major problem with this approach is the rapid clogging of the filter bed, due to solids carry-over from the sedimentation tank. This can be reduced by microscreening, but this imposes an additional maintenance chore. The sedimented solids from aerated lagoons should ideally be subjected to anaerobic digestion or placed on sludge-drying beds until it has stabilised.

Recommended Further Reading

CHAPTER 1

Books

Bacterial Indicators/Health Hazards Associated with Water (1977) Special Technical Publication 635, American Society for Testing and Materials.

Eutrophication of Lakes and Reservoirs in Warm Climates (1988) Environmental Health Series, vol. 30, World Health Organisation.

Feachem, R. G., Bradley, D. J., Garelick, H. and Mara, D. D. (1983) *Sanitation and Disease—Health Aspects of Excreta and Wastewater Management*, Wiley.

Hart, C. W. and Fuller, S. L. H. (1974) *Pollution Ecology of Freshwater Invertebrates*, Academic Press.

Hutton, L. (1983) *Field Testing of Water in Developing Countries*, Water Research Centre.

Klein, L. (1957) *Aspects of River Pollution*, Butterworths.

Koning, H. W. (1987) *Setting Environmental Standards—Guidelines for Decision Making*, World Health Organisation.

Lamb, J. C. (1985) *Water Quality and its Control*, Wiley.

Pike, E. B. (1975) Aerobic bacteria, in *Ecological Aspects of Used Water Treatment*, vol. 1 (eds C. R. Curds and H. A. Hawkes), Academic Press.

Ravera, O. (1979) *Biological Aspects of Freshwater Pollution*, Pergamon Press.

Velz, C. J. (1984) *Applied Stream Sanitation*, Wiley.

Winblad, U. and Kilama, W. (1985) *Sanitation Without Water*, Maunillans.

References

McBride, G. B. (1982) Nomographs for rapid solutions for the Streeter–Phelps equations, *Journal of the Water Pollution Control Federation*, **54**, 378–84.

Negulesco, M. and Rojanski, V. (1962) Recent research to determine reaeration coefficient, *Water Research*, **3**, 189–202.

Painter, H. A. (1958) Some characteristics of a domestic sewage, *Water and Waste Treatment Journal*, **6**, 496–504.

Royal Commission on Sewage Disposal (1912) *Standards and Tests for Sewage and Sewage Effluents Discharging into Rivers and Streams*, Eighth Report, HMSO.

Streeter, H. W. and Phelps, E. B. (1925) A study of the pollution and natural purification of the Ohio river, *U. S. Public Health Bulletin*, No. 146.

Warn, A. E. and Brew, J. S. (1980) Mass balance, *Water Research*, **14**, 1427–34.

Woodiwiss, F. S. (1964) The biological system of stream classification used by the River Trent Board, *Chemistry and Industry*, **11**, 443–47.

CHAPTER 2

Books

Barnes, D. Forster, C. and Johnstone, D. (1983) *Oxidation Ditches in Wastewater Treatment*, Pitman Books.

Curds, C. R. and Hawkes, H. A. (eds) (1983) *Biological Activities and Treatment Processes*, Ecological Aspects of used Water Treatment, vol. 2, Academic Press.

Manuals of British Practice in Water Pollution Control (1973–88) *Primary Sedimentation* (1973), *Preliminary Processes* (1984), *Activated Sludge* (1987), *Biological Filtration* (1988), Institute of Water annd Environmental Management.

Tebbutt, T. H. Y. (1971) *Principles of Water Quality Control*, Pergamon Press.

Water Pollution Control (1986) Module 3, Wastewater treatment Work Unit 1—The primary treatment of wastewater, HMSO.

References

Backhurst, J. R., Harker, J. H. and Kaul, S. N. (1988) The characteristics of aerators for the oxygenation of activated sludge *Asian Environment*, **14**, 65–72.

Chambers, B. and Jones, G. L. (1985) Energy savings by fine bubble aeration, *Water Pollution Control*, **84**, 70.

Clough, G. F. G. (1974) Physical characteristics of mechanical aerators, *Water Pollution Control*, **73**, 564.

Collins, O. C. and Elder, M. D. (1980) Experience in operating a deep shaft process, *Water Pollution Control*, **79**, 272.

Hitdlebaugh, J. A. and Miller, R. D. (1981) Operational problems with rotating biological contactors, *Journal of the Water Pollution Control Federation*, **53**, 1283–93.

Kalinske, A. A. (1976) A comparision of air and oxygen activated sludge systems, *Journal of the Water Pollution Control Federation*, **48**, 2472.

Morales, L. and Reinhart, D. (1984) Full-scale evaluation of aerated grit chambers, *Journal of the Water Pollution Control Federation*, **56**, 337–43.

Pike, E. B. (1978) The design of percolating filters and rotary biological contactors, including details of international practice, *Water Research Centre Technical Report TR 93*.

Pike, E. B., Carlton-Smith, C. H., Evans, R. H. and Harrington, D. W. (1982) Performance of rotating biological contactors under field conditions, *Water Pollution Control*, **81**, 10.

Upton, J. Norman, M. and Crabtree, H. E. (1985) The Vitox process: its oxygen utilisation efficiency and the optimal mode of deployment in the uprating context, *La Tribune CEBEDEAU*, **501** (38), 29–45.

CHAPTER 3

Books

Clarifier Design (1983) Manual of Practice for Water Pollution Control FD-8, Water Pollution Control Federation.

Forster, C. F. (1985) *Biotechnology and Wastewater Treatment*, Cambridge Studies in Biotechnology, vol. 2, Cambridge University Press.

Ganczarczyk, J. J. (1983) *Activated Sludge Process—Theory and Practice*, Marcel Dekker.

Gibbon, D. L. (1974) *Aeration of Activated Sludge in Sewage Treatment*, Pergamon.

James, A. (1984) *An Introduction to Water Quality Modelling*, Wiley.

Levenspiel, O. (1967) *Chemical Reaction Engineering*, Wiley.

Measurement of Oxygen Transfer in Clean Water (1985) ASCE Standard, American Society of Civil Engineers.

Winkler, M. (1981) *Biological Treatment of Wastewater*, Ellis Horwood.

References

Bode, H. and Seyfried, C. F. (1984) Mixing and detention time distribution in activated sludge tanks, *12th IAWPRC Biennial International Conference*, pp. 197–208.

Catunda, P. F. C. and van Haandel, A. C. (1986) Activated sludge settlers: design and optimization, *Water Science and Technology*, **19**, 613–23.

Handley, J. (1974) Sedimentation: an introduction to solids flux theory, *Water Pollution Control*, **73**, 230–40.

Mueller, J. A. and Boyle, W. C. (1988) Oxygen transfer under process conditions, *Journal of the Water Pollution Control Federation*, **60**, 332–41.

Smart, P. L. and Laidlaw, I. M. S. (1977) An evaluation of some fluorescent dyes for water tracing, *Water Resources Research*, **13**, 15–33.

Tebbutt, T. H. Y. and Christoulas, D. G. (1982) Performance relationships for primary sedimentation, *Water Research*, **16**, 347–56.

Tewari, P. K. and Bewtra, J. K. (1982) Alpha and beta factors for domestic wastewater, *Journal of the Water Pollution Control Federation*, **54**, 1281–87.

Thirumurthi, D. (1969) Design principles of waste stabilisation ponds, *Proceedings of the American Society of Civil Engineers, Sanitary Engineering Division*, **95**, SA2.

Tomlinson, E. J. and Chambers, B. (1979) The effects of longitudinal mixing on the settleability of activated sludge, *Water Research Centre Technical Report TR 122*.

White, M. J. D. (1975) Settling of activated sludge, *Water Research Centre Technical Report TR 11*.

White, M. J. D. (1976) Design and control of secondary settlement tanks, *Water Pollution Control*, **75**, 457–67.

Wimpenny, J. W. T. (1977) Water tracing, in *Treatment of Industrial Effluents* (ed. A. G. Calley Hodder & Stoughton, pp. 346–75.

CHAPTER 4

Books

E. G. Bellinger (1980) *A Key to Common British Algae*, Institute of Water Engineers and Scientists.

Curds, C. R. (1969) *An Illustrated Key to the British Ciliated Protozoa Commonly Found in Activated Sludge*, Water Pollution Research Technical Paper No. 12.

Drasser, B. S. and Barrow, P. A. (1985) *Intestinal Microbiology*, Aspects of Microbiology vol. 10, Van Nostrand Reinhold (UK).

Gaudy, A. F. and Gaudy, E. T. (1980) *Microbiology for Environmental Scientists and Engineers*, McGraw-Hill.

Mara, D. D. (1974) *Bacteriology for Sanitary Engineers*, Churchill Livingstone.

Mudrack, K. and Kunst, S. (1986) *Biology of Sewage Treatment and Water Pollution Control*, Ellis Horwood.

Rao, V. C. and Melnick, J. L. (1986) *Environmental Virology*, Aspects of Microbiology, vol. 13, Van Nostrand Reinhold (UK).

Rheinheimer, G. (1985) *Aquatic Microbiology*, Wiley.

Sterritt, R. M. and Lester, J. N. (1988) *Microbiology for Environmental and Public Health Engineers*, E. & F. N. Spon.

Wilkinson, J. F. (1975) *Introduction to Microbiology*, Blackwell Scientific.

References

Amin, P. M. and Ganapati, S. V. (1967) Occurrence of *Zoogloea* colonies and protozoans at different stages of sewage purification, *Applied Microbiology*, **15**, 17–21.

Curds, C. R. (1973) The role of protozoa in the activated sludge process, *American Zoology*, **13**, 161–9.

Curds, C. R., Cockburn, A. and Vandyke, J. M. (1968) An experimental study of the role of ciliated protozoa in the activated sludge process, *Water Pollution Control*, **67**, 312–29.

Harris, R. H. and Mitchell, R. (1973) The role of polymers in microbial aggregration, *Annual Review of Microbiology*, **27**, 27–50.

Kucnerowicz, F. and Verstraete, W. (1983) Evolution of microbial communities in the activated sludge process, *Water Research*, **17**, 1275–9.

Taber, W. A. (1976) Wastewater microbiology, *Annual Review of Microbiology*, **30**, 263–77.

Williams, T. M. and Unz, R. F. (1983) Environmental distribution of *Zoogloea* strains, *Water Research*, **17**, 779–87.

CHAPTER 5

Books

Anderson, J. W. (1980) *Bioenergetics of Autotrophs and Heterotrophs*, Arnold.

Dawes, E. A. (1980) *Quantitative Problems in Biochemistry*, 6th edn, Longman.

Gaudy, A. F. and Gaudy, E. T. (1981) *Microbiology for Environmental Engineers and Scientists*, McGraw-Hill.

Gottschalk, G. (1986) *Bacterial Metabolism*, Springer-Verlag.

Hall, D. O. and Rao, K. K. (1987) *Photosynthesis*, Arnold.

Lynch, J. M. and Poole, N. J. (1978) *Microbial Ecology, a Conceptual Approach*, Blackwell Scientific.

Mandelstam, J., McQuillen, K., and Dawes, I. (1968) *Biochemistry of Bacterial Growth*, Blackwell Scientific.

Sterritt, R. M. and Lester, J. N. (1988) *Microbiology for Environmental and Public Health Engineers*, E. & F. N. Spon.

Tribe, M. and Whittaker, P. (1972) *Chloroplasts and Mitochondria*, Arnold.

References

Gottschalk, G. and Andreesen, J. R. (1979) Energy metabolism in anaerobes, *Microbial Biochemistry*, **21**, 85–115.

Haddock, B. A. and Jones, C. W. (1977) Bacterial respiration, *Bacteriology Reviews*, **41**, 47–99.

Konings, W. N. and Veldkamp, H. (1983) Energy transduction and solute transport mechanisms in relation to environments occupied by microorganisms, *Microbes in their Natural Environments, Society for General Microbiology Symposium 34*, Cambridge University Press.

Mitchell, P. (1972) Chemiosmotic coupling and energy transduction: a logical development of biochemical knowledge, *Bioenergetics*, **3**, 5–24.

Morris, J. G. (1975) The physiology of obligate anaerobiosis, *Advances in Microbial Physiology*, **12**, 169–246.

Slater, E. C. (1981) The discovery of oxidative phosphorylation, *Trends in Biochemical Sciences*, **6**, 226–7.

Thauer, R. K., Jungermann, K. and Decker, K. (1977) Energy conservation in chemotrophic anaerobic bacteria, *Bacteriological Reviews*, **41**, 100–80.

CHAPTER 6

Books

Gaudy, A. F. and Gaudy, E. T. (1980) *Microbiology for Environmental Engineers and Scientists*, McGraw-Hill.

Lynch, J. M. and Poole, N. J. (1979) *Microbial Ecology: A Conceptual Approach*, Blackwell Scientific.

References

Curds, C. R. and Bazin, M. J. (1977) Protozoan predation in batch and continuous culture, *Advances in Aquatic Microbiology*, **1**, 115–75.

Daigger, G. T. and Grady, C. P. L. (1982) The dynamics of microbial growth on soluble substrates—a unifying theory, *Water Research*, **16**, 365–82.

Mallory, L. M., Yuk, C. S., Liang, L. N. and Alexander, M. (1983) Alternative prey: a mechanism for elimination of bacterial species by protozoa, *Applied and Environmental Microbiology*, **46**, 1073–9.

Monod, J. (1949) The growth of bacterial populations *Annual Review of Microbiology*, **3**, 371–94.

Payne, W. J. (1970) Energy yields and growth of heterotrophs, *Annual Review of Microbiology*, **24**, 17–51.

Stouthammer, A., and Bettenhaussen, C. (1973) Utilization of energy for growth and maintenance in continuous batch culture of microorganisms, *Biochimica Biophysica Acta*, **30**, 53–70.

Tempest, D. W. (1968) The continuous cultivation of microrganisms 1. Theory of the chemostat, *Methods in Microbiology*, **2**, 259–79.

CHAPTER 7

Books

Ganczarczyk, J. J. (1983) *Activated Sludge Process—Theory and Practice*, Marcel Dekker.

James, A. (1984) *An Introduction to Water Quality Modelling*, Wiley.

Nicoll, E. H. (1988) *Small Water Pollution Control Works: Design and Practice*, Ellis Horwood.

Spearing, B. W. (ed.) (1987) *STOM User Manual and Description*, 2nd edn.

Sundstrom, D. and Klei, H. E. (1979) *Wastewater Treatment*, Prentice-Hall.

Wilson, F. (1981) *Design Calculations in Wastewater Treatment*, E. & F. N. Spon.

Winkler, M. (1981) *Biological Treatment of Wastewater*, Ellis Horwood.

References

Baker, J. M. and Graves, Q. B. (1968) Recent approaches for trickling filter design, *Journal of the Sanitary Engineering Division, ASCE*, **94** (SA1), 65–84.

Chambers, B. and Jones, G. L. (1988) Optimisation and uprating of activated sludge plants by efficient process design, *Water Science and Technology*, **20**, 121–32.

Eckenfelder, W. W. Jr (1961) Trickling filtration design and performance, *Journal of the Sanitary Engineering Division, ASCE*, **87** (SA4), 33–45.

Goodman, B. L. and Englande, J. Jr (1974) A unified model of the activated sludge process, *Journal of the Water Pollution Control Federation*, **46**, 312–32.

IAWPRC Task Group on Mathematical Modelling for Design and Operation of Biological Wastewater Treatment (1987) Final report: Activated sludge model, *IAWPRC Scientific and Technical Reports*, No. 1.

Lawrence, A. W. and McCarty, P. L. (1970) Unified basis for biological treatment design and operation, *Journal of the Sanitary Engineering Division, ASCE*, **96** (SA3), 757–78.

McKinney, R. E. (1962) Mathematics of complete-mixing activated sludge, *Journal of the Sanitary Engineering Division, ASCE*, **88**(SA3), 87–113.

Marais, G. v. R. and Ekama, G. A. (1976) The activated sludge process, Part I—Steady state behaviour *Water SA*, **2**, 163–200.

Randall, C. W., Benefield, L. D. and Buth, D. (1982) The effects of temperature on the biochemical reaction rates of the activated sludge process, *Water Science and Technology*, **14**, 413–430.

Spearing, B. W. (1987) Sewage treatment optimization model—STOM—the sewage works in a personal computer, *Proceedings of the Institution of Civil Engineers*, **82**, 1145—64.

Velz, C. J. (1948) A basic law for the performance of biological filters *Sewage Works*, **20**, 607–17.

CHAPTER 8

Books

Barnes, D. and Bliss, P. J. (1983) *Biological Control of Nitrogen in Wastewater Treatment*, E. & F. N. Spon.

Cole, J. A. and Ferguson, S. J. (1988) *The Nitrogen and Sulphur Cycles*, Society for General Microbiology Symposium 42, Cambridge University Press.

Nutrient Control (1983) Manual of Practice FD-7, Facilities design, Water Pollution Control Federation.

Payne, W. J. (1981) *Denitrification*, Wiley.

J. I. Prosser (1986) *Nitrification*, Society for General Microbiology Special Publication, vol. 20, IRL Press.

Wanielista, M. P. and Eckenfelder, W. W. Jr (1978) *Biological Nutrient Removal*, Advances in Water and Wastewater Treatment, Ann Arbor Science.

References

Argaman, Y., and Brenner, A. (1986) Single-sludge nitrogen removal: modelling and experimental results, *Journal of the Water Pollution Control Federation*, **58**, 853–60.

Comeau, Y., Rabionwitz, B., Hall, K. J. and Oldham, W. K. (1987) Phosphate release and uptake in enhanced biological phosphorus removal from wastewater, *Journal of the Water Pollution Control Federation*, **59**, 707–15.

Downing, A. L., Painter, H. A., and Knowles, G. (1965) Nitrification in the activated sludge process, *Journal of the Proceedings of the Institute of Sewage Purification*, **2**, 130–142.

Knowles, R. (1982) Denitrification, *Microbiological Reviews*, **46**, 43–70.

Painter H. A. and Loveless, J. E. (1983) Effect of temperature and pH value on the growth rate constants of nitrifying bacteria in the activated sludge process, *Water Research*, **17**, 237–248.

Parker, D. S. and Richards, T. (1986) Nitrification in trickling filters, *Journal of the Water Pollution Control Federation*, **58**, 896–902.

Skrinde, J. R. and Bhagat, S. K. (1982) Industrial wastes as carbon sources in biological denitrification, *Journal of the Water Pollution Control Federation*, **54**, 370–7.

Teichgraeber, B. (1988) Operational problems with two-stage activated sludge plants with nitrification, *Water Supply*, **6**, 125–32.

Upton, J. and Cartwright, D. (1984) Basic design criteria and operating experience of a large nitrifying filter, *Water Pollution Control*, **83**, 340–52.

Wiechers, H. N. S. and Heynike, J. J. C. (1986) Sources of phosphorus which give rise to eutrophication in South African waters, *Water SA*, **12**, 99–102.

Wentzel, M. C., Lotter, L. H., Loewenthal, R. E. and Marais, G.V.R. (1986) Metabolic behaviour of *Acinetobacter* spp. in enhanced biological phosphorus removal-a biochemical model *Water SA*, **12**, 209–224.

CHAPTER 9

Books

Chambers, B. and Tomlinson, E. J. (1982) *Bulking of Activated Sludges: Preventative and Remedial Measures*, Ellis Horwood.

Effluent Variability (1976) Progress in Water Technology, vol. 8, Pergamon Press.

Eikelboom, D. H. and van Buijsen, H. J. J. (1981) *Microscopic Sludge Investigation Manual*, TNO Research Institute for Environmental Hygiene, The Netherlands.

Jenkins, D., Richard, M. G. and Daigger, G. T. (1986) *Manual on the Causes and Control of Activated Sludge Bulking and Foaming*, Water Research Commission, Pretoria, South Africa.

Junkins, R., Deeny, K. and Eckbot, F. T. (1983) *The Activated Sludge Process: Fundamentals of Operation*, Ann Arbor Science.

Nicoll, E. H. (1988) *Small Water Pollution Control Works Design and Practice*, Ellis Horwood.

Operation and Management of Small Sewage Works (1971) HMSO.

Operations and Maintenance (1987) Activated Sludge Manual of Practice OM-9, Water Pollution Control Federation.

References

Chiesa, S. C. and Irvine, R. L. (1985) Growth and control of filamentous microbes in activated sludge: an integrated hypothesis, *Water Research*, **19**, 471–9.

Chudoba, J., Cech, J. S. and Grau, P. (1985) Control of activated sludge filamentous bulking—experimental verification of a kinetic selection theory, *Water Research*, **19**, 191–6.

Curds, C. R. and Cockburn, A. (1970) Protozoa in biological sewage treatment processes—II. Protozoa as indicators in the activated sludge process, *Water Research*, **4**, 237–44.

Donald, M., Riddell, R., Lee, J. S. and Wilson, T. E. (1983) Method for estimating the capacity of an activated sludge plant, *Journal of the Water Pollution Control Federation*, **55**, 360–8.

Eikelboom, D. H. (1975) Filamentous organisms observed in bulking activated sludge, *Water Research*, **9**, 365–82.

Forster, C. F. and Dallas-Newton, J. (1980) Activated sludge settlement—some suppositions and suggestions. *Water Pollution Control*, **79**, 338–51.

Forster, C. F. (1985) Factors involved in the settlement of activated sludge—I. Nutrients and surface polymers, *Water Research*, **19**, 1259–64.

Harbott, B. J. and Penney, C. J. (1983) The efficiency of insecticide treatment of flies on biological filters, *Water Pollution Control*, **82**, 571–81.

Lovett, D. A., Kavanagh, B. V. and Herbert, L. S. (1983) Effect of sludge age and substrate composition on the settling and dewatering characteristics of activated sludge, *Water Research*, **11**, 1511–15.

Niku, S. and Schroeder, E. D. (1981) Factors affecting effluent variability from activated sludge processes, *Journal of the Water Pollution Control Federation*, **53**, 546–59.

Sezgin, M. (1982) Variation of sludge volume index with activated sludge characteristics, *Water Research*, **16**, 83–8.

White, M. J. D. (1975) Settling of activated sludge, *Water Research Centre Technical Report TR11*.

CHAPTER 10

Books

Curds, C. R. and Hawkes, H. A. (1983) *Ecological Aspects of Used-water Treatment*, Biological Activities and Treatment Processes, vol. 2, Academic Press.

Mara, D. D. (1978) *Sewage Treatment in Hot Climates*, Wiley.

Mara, D. D. and Marecos Do Monte, M. H. (1987) *Waste Stabilisation Ponds*, Water Science and Technology, vol. 19, Pergamon Press.

Mara, D. D. and Pearson, H. W. (1987) *Waste Stabilization Ponds*, Design Manual for Mediterranean Europe, World Health Organisation.

Middlebrooks, E. J., Middlebrooks, C. H., Reynolds, J. H., Watters, G. Z., Reed, S. C. and George, D. B. (1982) *Wastewater Stabilisation Lagoon Design, Performance and Upgrading*, Collier-Macmillan.

Pescod, M. B. and Arar, A. (1988) *Treatment and Use of Sewage Effluent for Irrigation*, Butterworths.

Wastewater Stabilization Ponds—Principles of Planning and Practice (1987) WHO EMRO Technical Publication No. 10, World Health Organisation.

References

Arthur, J. P. (1983) Notes on the design and operation of waste stabilization ponds in warm climates of developing countries, *World Bank Technical Paper Number 7*, The World Bank.

Bartsch, E. H. and Randall, C. W. (1971) Aerated lagoons—a report on the state of the art, *Journal of the Water Pollution Control Federation*, **43**, 699–708.

Health aspects of wastewater and excreta use in agriculture and aquaculture: The Engelberg Report (1985), *IRCWD News*, **23**, 11–18.

Mara, D. D. and Pearson, H. W. (1986) Artificial freshwater environments: waste stabilization ponds, *Biotechnology*, **8**, 177–206, VCH Verlagsgesellschaft, Weinheim.

Mara, D. D. and Pearson, H. W. (1987) Designing against pond horrors, *World Water*, **10**, 17–21.

Marais, G.v.R. (1974) Faecal bacterial kinetics in stabilisation ponds, *Journal of the Environmental Engineering Division, Proceedings of the American Society of Civil Engineers*, **100** (EE1), 119–39.

Pano, A. and Middlebrooks, E. J. (1982) Ammonia nitrogen removal in facultative wastewater stabilisation ponds, *Journal of the Water Pollution Control Federation*, **54**, 344–51.

Panicker, P. V. R. C. and Krishnamoorthis, K. P. (1981) Parasite egg and cyst reduction in oxidation ditches and aerated lagoons, *Journal of the Water Pollution Control Federation*, **53**, 1413–19.

Pearson, H. W. (1987) Applications of algae in sewage treatment processes, *Microbial Technology in the Developing World*, pp. 260–288 Oxford University Press.

Rich, L. G. (1982) Design approach to dual-power aerated lagoons, *Journal of the Environmental Engineering Division, Proceedings of the American Society of Civil Engineers*, **108**, 532–48.

Stentiford, E. I. (1983) Oxidation ditches, aerated lagoons and waste stabilisation ponds, *Comprehensive Biotechnology*, pp. 913–37. Pergamon Press.

List of Most Common Abbreviations

CHAPTER 1

D	Dissolved oxygen deficit (mg O_2/l)
k_1	BOD removal constant (t^{-1})
k_2	Reaeration constant (t^{-1})
L	BOD of a wastewater or river water (mg O_2/l)
Q	Flow rate (m^3/d)
t_c	Time of the critical oxygen deficit (t)
Y	Amount of BOD exerted (mg O_2/l)
R	Temperature correction coefficient

CHAPTER 3

A	Area of clarifier (m^2)
C	Dimensionless concentration
C_s	Saturation oxygen concentration (mg O_2/l)
D	Longitudinal dispersion coefficient
D_L	Diffusion coefficient (area/time)
G	Solids flux (kg/m^2 h)
k	Reaction rate constant (units variable)
K_L	Gas transfer coefficient (length/time)
$K_L a$	Mass transfer coefficient (t^{-1})
N	Number of tanks in series
Q	Flow rate (m^3/s)
R_O	Mixed liquor oxygen uptake rate (mg O_2/l)
t	Hydraulic retention time (t)
U	Fluid velocity (m/s)
V	Volume (m^3)
v	Particle settling velocity (length/time)
X	Solids concentration (mg/l)
y_f	Thickness of the gas/liquid interface (length)
z	Dimensionless time
θ	Dimensionless length/temperature correction coefficient
σ^2	Variance
δ	Dispersion number

CHAPTER 6

k_d	Decay coefficient (t^{-1})
k_s	Half-saturation coefficient (mg/l)
S	Substrate concentration (mg/l)
t_d	Bacterial doubling time (t)
q	Substrate utilisation rate (t^{-1})
X	Bacterial cell mass (mg/l)
Y	Yield coefficient (mg cells/mg substrate)
μ	Specific growth rate (t^{-1})
μ_m	Maximum specific growth rate (t^{-1})
θ	Solids retention time (t)

CHAPTER 7

A	Filter area (m^2)
D	Filter depth (m)
Q	Wastewater flow rate (m^3/d)
Q_r	Recycle flow rate (m^3/d)
R	Recycle ratio
S_e	Effluent BOD (mg/l)
S_0	Influent BOD (mg/l)
t	Hydraulic retention time (t)
V	Reactor volume (m^3)
X	Mixed liquor suspended solids (mg/l)
X_e	Effluent suspended solids (mg/l)
X_r	Recycle suspended solids (mg/l)
θ	Sludge age (d)

CHAPTER 8

DO	Dissolved oxygen concentration (mg/l)
K_{DO}	Half-saturation constant for dissolved oxygen (mg/l)
K_N	Ammonium half-saturation constant (mg/l)
N_e	Effluent TKN (mg/l)
N_0	Influent TKN (mg/l)
$(q)_{DN}$	Specific denitrification rate (t^{-1})
$\mu_{N,max}$	Maximum specific growth rate for nitrifying bacteria (t^{-1})

CHAPTER 10

A_f	Pond mid-depth area (m^2)
D_f	Pond mean depth (m)
k_b	first-order faecal coliform removal constant (d^{-1})

N_e	Number of faecal coliforms in effluent (/100 ml)
N_i	Number of faecal coliforms in influent (/100 ml)
t	Hydraulic retention time (t)
λ_s	Permissible surface loading (kg/ha d)
λ_v	Volumetric organic loading rate (g/m^3 d)

Index